Contents

INTRODUCTION

1	Basic Components	9
2	Installing Raspbian	12
3	Configuring Raspberry Pi	17
4	Raspi-config	23
5	Headless Raspberry Pi Setup	29
6	Programming the Raspberry Pi	34
7	LCD Alphanumeric Display	38
8	DHT11 Humidity and Temperature Sensor	47
9	Mini Weather Station - Part 1	53
10	Mini weather Station - Part 2	61
11	GPIO Connector Installation	67

12	GPIOZero Python Library	75
13	LED Control With GPIOZero - Part 1	84
14	LED Control With GPIOZero - Part 2	92
15	LED Control With GPIOZero - Part 3	97
16	Digital Input: Pull-Up and Pull-Down	100
17	Button with GPIOZero - Part 1	104
18	Button with GPIOZero - Part 2	109
19	GPIOZero Relay and Library	113
20	Led Brightness Control With GPIOZero and PWM	120
21	Led Bar with GPIOZero	126
22	RGB LED with GPIOZero	130
23	PIR Motion Sensor with GPIOZero	138
24	LDR Light sensor with GPIOZero	145
25	Raspberry PI 4: More Power, More RAM and 4K Video	152
26	20 Raspberry Pi Project Ideas	159

CONCLUSION

RASPBERRY PI 4

Raspberry Pi 4

A Comprehensive Step-by-Step Guide to Using Raspbian to Create Amazing Projects and Improve Your Programming Skills with the Latest Version of Raspberry Pi

LIAM CLARK

Liam Clark

© Copyright 2021 - All rights reserved.

The content in this book may not be reproduced, duplicated, or transmitted without direct written permission from the author or the publisher.

Under no circumstances will any blame or legal responsibility be held against the publisher, or author, for any damages, reparation, or monetary loss due to the information contained within this book. Either directly or indirectly.

Legal Notice:

This book is copyright protected. This book is only for personal use. You cannot amend, distribute, sell, use, quote, or paraphrase any part, or the content within this book, without the consent of the author or publisher.

Disclaimer Notice:

Please note the information contained within this document is for educational and entertainment purposes only. All effort has been executed to present accurate, up-to-date, and reliable, complete information. No warranties of any kind are declared or implied. Readers acknowledge that the author is not engaging in the rendering of legal, financial, medical, or professional advice. The content within this book has been derived from various sources. Please consult a licensed professional before attempting any techniques outlined in this book.

By reading this document, the reader agrees that under no circumstances is the author responsible for any losses, direct or indirect, which are incurred as a result of the use of the information contained within this document, including, but not limited to—errors, omissions, or inaccuracies.

Introduction

Raspberry Pi is a single-board computer, much used by makers in DIY. In this guide, we will see how to configure it and use it to create applications of various types. Starting from the study of the board, we will proceed with the configuration and installation of the basic system, until we discover the potential offered by Raspberry Pi from the maker point of view.

What is a Raspberry Pi?

The Raspberry Pi is a single-board computer (a specific case of System on Chip, or SoC), with ARM CPU. This board was developed in the UK by the Raspberry Pi Foundation, and the first models were presented to the public during the winter of 2012. The idea behind the realization of Raspberry Pi is to allow the teaching of computer science and programming in schools, through the use of low-cost devices.

Raspberry Pi is available in different forms (form factor). Each available version has special features and stands out from the others for performance and/or capacity.

The different types of form factors are marked by a letter of the alphabet, possibly followed by a symbol (+). The latter distinguishes the models that have more resources than the equivalent models without the +, while sharing the form.

In this guide, we will see how to configure and use Raspberry Pi to create applications of various types. Starting with the study of the board, we will proceed with the configuration and installation of the basic system for daily use as a mini-computer. Once this is done, we will get to know Raspberry from another point of view, that is, as a device for the world of makers, thus discovering the potential that the board offers by using it in contexts different from that of a simple mini-computer.

First, here is a brief description of the various models currently available.

1. **Raspberry Pi 1 Model A+.**

   It is the low-cost variant of the Raspberry Pi. Compared to the basic A version, it has more GPIO pins (which we will talk about later in this guide). The GPIO interface contains 40 pins, and the first 26 maintain the same pinout as the previous Raspberry Pi A and Raspberry Pi B models. The old version of the socket that houses the SD card has been replaced by a push-push system for micro SD. Power consumption has been reduced by about 1 Watt. The audio quality has been improved and the form factor has been made smaller and more organized.

2. **Raspberry Pi 1 Model B+.**

   It is the final revision of the original Raspberry Pi. This version of Raspberry Pi replaced the older Model B in July 2014, to be later replaced by the Raspberry Pi 2 Model B. The GPIO interface contains 40 pins, and the first 26 maintain the same pinout as the previous models. The USB ports have been increased and the board includes four, two more than the previous Model B. The push-push system for reading the micro SD has been maintained and power consumption has been reduced even in this model by about 1 watt. The audio quality has been improved and the form factor has been reorganized to accommodate the new USB ports.

3. **Raspberry Pi 2 Model B.**
   It represents the second generation of Raspberry Pi. This board model replaces the first generation of Raspberry Pi in February 2015. The main differences from the previous version are:
   1. Cortex-A7 quad-core processor, with a frequency of 900MHz;
   2. 1 GB RAM.

4. **Raspberry Pi 3 Model B.**
   This Raspberry model is the evolution of its predecessor Raspberry Pi 2 B, which improves connectivity. The push-push system for reading the micro SD has been maintained and the power supply has been enhanced, supporting up to 2.5 A. The third generation Raspberry Pi was released in February 2016 and is equipped with:
   1. Quad-core Broadcom BCM2837 64-bit quad-core processor with 1.2 GHz frequency;
   2. 1 GB RAM;
   3. A low energy bluetooth and wireless connectivity module BCM43438 (BLE);
   4. A 40-pin GPIO interface;
   5. Four USB 2.0 ports;
   6. A 4-pole stereo output with composite video port;
   7. An HDMI port as video-audio output;
   8. A CSI port to connect the Raspberry Pi Camera;
   9. A DSI port to connect the Raspberry Pi touchscreen display.

5. **Raspberry Pi 3 Model B+.**
   Released in March 2017, it's equipped with:
   1. Broadcom BCM2837B0, Cortex-A53 (ARMv8) 64-bit SoC processor with 1.4GHz frequency;
   2. 1 GB RAM;

3. Connectivity - 2.4GHz and 5GHz IEEE 802.11.b/g/n/ac wireless LAN, Bluetooth 4.2, BLE;
4. Gigabit Ethernet over USB 2.0;
5. A 40-pin GPIO interface;
6. An HDMI port as video-audio output;
7. Four USB 2.0 ports;
8. A CSI port to connect the Raspberry Pi Camera;
9. A DSI port to connect the Raspberry Pi touchscreen display;
10. A 4-pole stereo output with composite video port;
11. A micro SD port to load the operating system and store data;
12. 5V/2.5A DC power supply;
13. Power-over-Ethernet (PoE) support (requires PoE HAT module).

6. **Raspberry Pi Zero.**
   It is exactly half the size of the A+ model, but twice as useful. Its features are:
   1. ASngle-core processor, with 1GHz frequency;
   2. 512 MB RAM;
   3. A mini-HDMI port as digital audio/video output;
   4. A micro-USB port for power supply;
   5. A micro-USB OTG port to connect various types of devices;
   6. A 40-pin grid as GPIO interface;
   7. Headers for reset and composite video output;
   8. A CSI connector for the camera.

7. **Raspberry Pi Zero W.**
   This model extends the Pi Zero family. Raspberry Pi Zero W was launched in February 2017 adding to its predecessor modules for connectivity, such as Bluetooth Low Energy 4.1 and Wireless 802.11 b/g/n. The Pi Zero W has

the same features as the Pi Zero described above.

8. **Raspberry Pi 4**.
   It is, to date, the latest version of the single-board computer produced by the Raspberry Foundation.
   The Raspberry Pi 4 Model B, in 1GB, 2GB, and 4GB memory configurations respectively, is released in June 2019, mounts a Broadcom BCM2711 quad-core 1.5 GHz quad-core, and offers many new features compared to its predecessors:
   1. RAM LPDDR4 (with versions of 1, 2, 4, and 8 gigabytes of RAM);
   2. Two micro HDMI ports with 4K support;
   3. 802.11ac dual-band WiFi module;
   4. Bluetooth V5.0;
   5. Two USB 3 + two USB 2 ports;
   6. Power supply via the new USB-C connector.

Raspberry supports several operating systems, mainly based on Unix Kernel (especially Linux/Debian) and RISC architectures. In this guide, we will use Raspbian, an operating system based on Linux Kernel and derived from Debian. In particular, we will see all the steps necessary to install your system and we will study the aspects and configurations needed to fully exploit the device. Before doing this, let's introduce more accurately the board, taking as a reference the Raspberry Pi 2 model. Although the latter is not the latest version available, the differences compared to Raspberry Pi 4 relate exclusively to performance and connectivity, so this choice will still allow you to generalize the contents of this guide to subsequent models.

The Board

Raspberry Pi 2 is presented as a small electronic board with integrated several modules to use the device as a simple computer or as a development board for the prototyping of intelligent modules for home automation, robotics or more generally for any automated system.

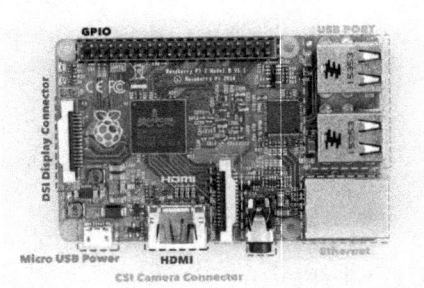

Raspberry Pi 2 B features a micro USB port for 5V power supply, an HDMI port for audio/video connection to monitors/TVs that support HDMI, a 3.5" jack for audio output and composite video output, a 10/100 Ethernet port, four USB 2.0 ports, and a player to insert the micro SD where the operating system that will manage the Raspberry Pi will be installed.

Viewed from above, the board features a DSI port to manage displays with DSI connectivity and a CSI port to use the Raspberry Pi Camera, a VideoCore IV 3D graphics core, and probably the most interesting part of this board: the 40-pin GPIO interface. It allows, as we will see, to develop projects of various types, typically based on the need to control actuators of various types, all with extreme simplicity.

Chapter 1

Basic Components

In this chapter, we will learn about some of the key components for starting and using the operating system on the Raspberry Pi. We will analyze step by step the components needed for the initial setup:

1. **SD Card**.
   SD cards are memory cards, whose purpose is generally to store digital data of all kinds. In particular, the micro SD card needed for the Raspberry Pi 2 can be equated to the same function that a hard drive for a computer has. On this card will be installed the operating system and will be organized and saved our data, such as pictures, documents, videos, images, scripts, or programs. Raspberry Foundation recommends using memory cards of at least 4GB and class 4, although we strongly recommend the use of cards with a capacity of at least 8GB and class 10, to make the best use of the card and avoid unnecessary freeze or crash of the mini-computer.

2. **HDMI cable or adapter**.
   To get the most out of your Raspberry as a computer,

you'll need the HDMI cable to connect it to a generic monitor or TV with an input. There are a number of inexpensive HDMI cables available on the market; if you do not have an HDMI cable or monitor/television with the required input, you can always opt to purchase an HDMI to VGA or HDMI to DVB-T adapter, so you can still use the Raspberry Pi. Note also that in this first part of the guide we will use the Raspberry Pi 2 B, but some form factors described in the first chapter provide an S-Video output: in this case, we will need an RCA video cable, to connect to the monitor. Using this method of connection between the screen and Raspberry Pi, the audio will have to be connected via a 3.5" jack cable.

3. **Keyboard and Mouse**.
Any USB mouse or keyboard generally works well. There is only one thing to consider: some USB keyboards and mice require a lot of electricity to work properly. If so, Raspberry Foundation recommends using an externally powered USB hub. This problem could occur if you also connect other USB devices to the board, such as Wi-Fi connection modules.

4. **Power supply unit**.
The power supply is one of the most important parts of the entire system. It must have well-defined characteristics, first of all, the type of power supply, which must be micro USB. The choice of the power supply must be calibrated to what we intend to do with our Raspberry, as all the power (or most of it) comes from it. A bad power supply will not only be the source of most problems but can seriously compromise the board. Generally, many smartphone chargers are suitable, but you should always check the label on the charger before connecting the board to

the power supply. The Raspberry requires a voltage of 5V and a current of at least 1.2-1.5A to work really well. If the power supply provides voltages below 5V, the Raspberry may not work properly or may malfunction. So be wary of any low-cost power supply if you care about the health of your mini computer.

5. **Ethernet cable**.

    This element, which for some might be optional, is actually indispensable if we want to install the system from scratch.

With these elements at our disposal, in the next chapter, we will proceed with the installation and configuration of the Raspbian operating system, available for Raspberry Pi and based on the best known Debian Linux distribution.

Chapter 2

Installing Raspbian

Using Raspberry Pi requires you to use a complete and working operating system. In the previous chapter, we saw the tools needed to use our Raspberry Pi. The initial setup, in addition to the board itself, includes components such as:

- **Power supply 5V 2A**, to power the board;
- **SD card**, where to install Raspbian operating system;
- **HDMI cable**, for connection to a monitor/TV;
- **Mouse and Keyboard**, to configure/manage/use Raspberry Pi;
- **Ethernet cable**, for Internet connection.

Raspbian

Raspbian is a Linux operating system, based on the Debian distribution for the ARM architecture. It is the most popular operating system to be used on Raspberry Pi, which is why in this chapter we will see how to install it.

GETTING RASPBIAN

To download Raspbian for Raspberry Pi, we must go to the official Raspberry Pi Foundation website.

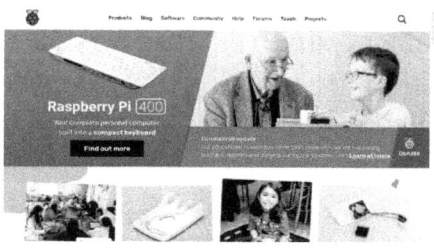

Raspberrypi.org homepage

Once you reach the home page, click on the **Software** tab at the top of the page.

By accessing the download page (www.raspberrypi.org/software/operating-systems/), you will be shown several tabs, each of which offers the possibility to download different types of operating systems supported by the board we have provided. As a first time, the advice is to download **NOOBS** (www.raspberrypi.org/downloads/noobs/): it is a simple operating system installer for the Raspberry Pi board, which allows you to choose between a number of operating systems (including Raspbian). You can also select the operating system you prefer from the download page, thus proceeding with the system set up on the SD card.

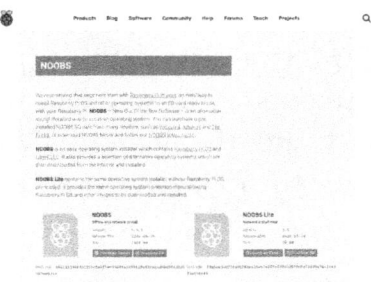

NOOBS download

From here you can download two versions of NOOBS: the *offline and network install* version and the *NOOBS Lite* version, i.e. network install only. In the classic version of NOOBS, Raspbian is already present in the .zipper archive that we will download by clicking on Download ZIP.

HOW TO INSTALL NOOBS ON YOUR SD CARD

Once the download is complete, we decompress the .zip file using any compressed file management program (for example 7zip on Windows or The Unarchiver on Mac OS).

We then insert the SD card that we will use on the Raspberry Pi into a card reader (now included on most PCs). This card must be formatted using the **FAT32** file system.

Once formatting is finished, we open the SD card and copy the newly decompressed files inside, taking care to copy the contents of the archive directly to the SD card, without creating additional folders.

After copying and pasting the entire NOOBS system, remove the SD card, remove it from the card reader and insert it into the slot on the Raspberry Pi.

STARTING NOOBS AND INSTALLATION

Completed this part, if no errors were made when copying files to the SD card, we connect to our Raspberry Pi the keyboard, mouse, Ethernet cable, HDMI cable, and power supply with the micro-USB connector in the appropriate ports, as discussed in the introductory lesson. Once the power supply is connected to our board, the monitor/TV connected to the Raspberry Pi will show the following interface:

RASPBERRY PI 4 ~ 15

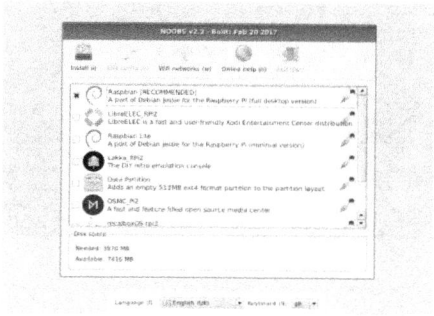

Click on the checkbox next to the operating system we have chosen for Raspberry Pi (Raspbian), and confirm our choice by clicking on Install. We will then be shown a confirmation message. We confirm by clicking on Yes and start the installation process.

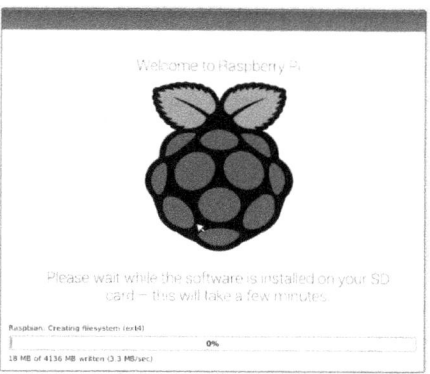

At the end of the installation, NOOBS will confirm that everything went well.

By clicking OK, our Raspberry Pi will automatically reboot and, if all went well, the screen will show the Raspbian bootstrap process and the loading of the operating system modules.

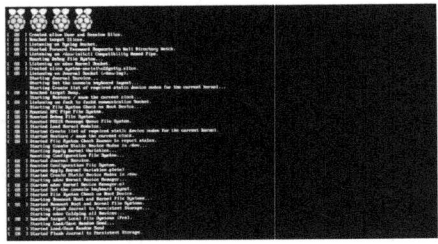

Raspbian Boot

After this preliminary system startup phase, we will finally have access to the Raspbian desktop, based on the new Pixel graphic manager.

Raspbian is now installed on our Raspberry Pi, and ready to be used or for all our purposes.

Chapter 3

Configuring Raspberry Pi

In the previous chapter, we saw how to install Raspbian on our Raspberry Pi. Generally, Raspbian keeps several services disabled, unused by those who want to use the board as if it is a small personal computer. To enable them, we can proceed in two different ways: by graphics (recommended for those who are not very familiar with the Linux terminal), or through appropriate commands executable from the terminal (less intuitive, but recommended for more experienced users). In this chapter, we will see how to proceed with the configuration by graphics, using the tools provided by Raspbian.

Once you have started Raspbian, click on the icon with the Raspberry Pi symbol at the top left of the desktop. The system will show a drop-down menu, from which we will have to select Preferences -> Raspberry Pi Configuration.

The following window will open:

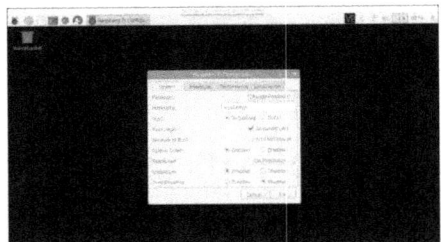

Configuring system settings on Raspbian

The previous image shows the configuration panel of the Raspberry Pi, divided by tabs into categories. In the first tab *System*, we can make those configurations related to the system. By clicking on the *Change Password...* button, the system will allow us to change the password of the user pi (i.e. the user-created by default by Raspbian), which by default is the *raspberry* string. The *Hostname* field, instead, allows you to set the network name of the board. This name is used by several protocols to identify and communicate with the device.

The third field, titled *Boot*, allows you to determine the system boot mode. By default, it is set to *To Desktop*, an option that allows you to launch the Pixel desktop environment at system startup; alternatively, you can select *To CLI*, making the system start the Command Line Interface. The fourth field, *Auto Login*, selected by default, logs in directly with the pi user; if disabled, we will have to specify which user to log in with. The *Splash Screen* option, set by default to *Enabled*, enables or disables the initial splash screen of Raspberry Pi. The other fields, *Network at Boot, Resolution, Underscan,* and *Pixel Doubling* allow more advanced configurations.

The second tab, Interfaces, allows the user to select which services or interfaces to enable on our device. The following image shows the basic configuration of our tab after installing the system.

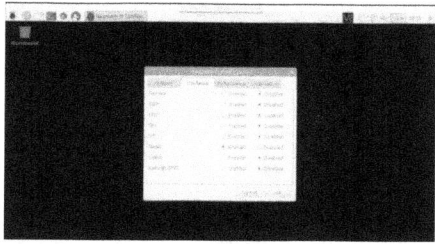

Configuration of services and interfaces on Raspbian

Each field allows you to enable or disable the related interface or service. The first field allows enabling the support of the CSI camera connector. The second field activates the SSH server when the Raspberry Pi starts.

We'll go further on using SSH to understand how powerful and useful this service is; for the moment, we'll just say that SSH support allows you to remotely manage the Raspberry Pi board connected to the network.

The third field activates the VNC server, also discussed later in this guide, which allows you to view the Raspberry Pi "screen" from any other computer equipped with a VNC client. The fourth field activates the SPI (Serial Peripheral Interface) synchronous serial communication interface, which is normally used for short-distance communication. The fifth field activates the I2C type communication interface, also a serial communication system used between integrated circuits. Serial, activated by default, enables the Raspberry Serial Interface. The seventh interface allows enabling the 1-wire communication service, while the eighth and last field allows starting the service that makes available the remote use of the GPIO interface.

At the moment we only enable those services that are useful to use Raspberry Pi at its best. The following image shows the configuration we use to continue with our guide.

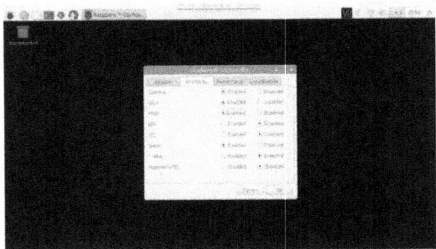

Enabling services and interfaces

Let's skip the Performance tab, from which we can change some settings related to the control and speed of the Raspberry Pi processor. It is advisable to avoid changing these options unless you know exactly what the consequences might be: you risk burning the device.

The last tab, *Localisation*, allows you to change the options related to the language and localization settings of the device:

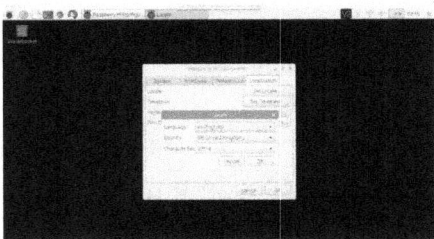

Raspbian Location Settings

By clicking on the *Set Local* button, associated with the first *Local* field, a new window will be displayed that allows you to set the language, the continent, and the character set that the system will use. Select the language you prefer in the Country field. As for the charset, set by default to UTF-8, we can select the Latin standard by clicking on ISO-8859-1 (you can select the set compatible with your language). Once the configuration of the localization settings is complete, we can configure the time zone (Timezone) by clicking on *Set Timezone*.

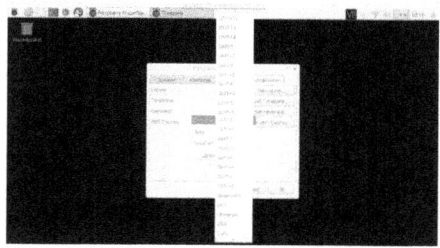

Time zone configuration

During installation, you can set the keyboard layout we are using with the Raspberry Pi; but if you forgot or skipped this step, you can fix it using the *Set Keyboard* button. In fact, a new window will open, called *Keyboard Layout*, from which we can select the country, and one of the many variants of keyboard layouts. In almost all cases, it will be sufficient to select the UK on the Country tab, and English on the Variant tab (as always, select the country and language of your preference).

The last configuration we have left is the one related to the wireless interface. With reference to the different European standards, to avoid potential problems related to disturbances or malfunctions due to incorrect configurations of the radio frequency interfaces, we can select which configuration Raspbian should use for WiFi. Then click on the Set WiFi Country button. In the window that will open, click on the drop-down menu and look for the item related to your country.

Confirming the change with the OK button, a dialog box will show to inform us that we will need to restart the system in order to make the changes effective. Click Yes and wait for the Raspberry Pi to restart. At the next boot, if nothing went wrong, we will have the system perfectly configured and totally in English.

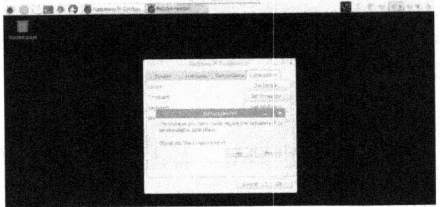

Reboot required at the end of the configuration

In the next chapter we will see how to make the same configurations using the terminal.

Chapter 4

Raspi-config

In the previous chapter, we saw how to configure our Raspberry Pi using the application that Raspbian provides, and which has a graphical interface. To enable the services that are disabled by default on Raspbian, you can configure the board using an alternative method to the application with a graphical interface, the **command-line configuration**, based on the use of the **raspi-config** tool.

Once Raspbian is started, click on the icon with the Terminal symbol.

A window will open in which a green text followed by a grey block describes the command line of the terminal:

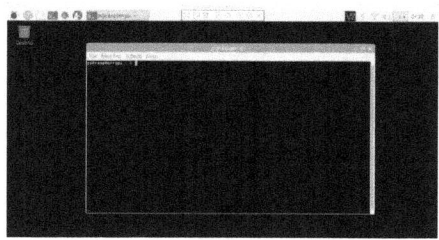

Terminal on Raspbian

The green text contains two important pieces of information. The colons (:) separates the information and define the user and hostname of the machine in the form *user@hostname*, followed by the current path. The ~ symbol represents the home folder of the user pi, i.e. */home/pi/*.

Let's see how to use *raspi-config* to configure Raspbian. Type the following into the terminal:

pi@raspberrypi:~$ sudo raspi-config

The *sudo* command allows us to obtain special administrator privileges when executing the command. Before executing the command, the system will ask for the password of the user pi, which by default is *raspberry*. When you type the password, the system will display an interface like this:

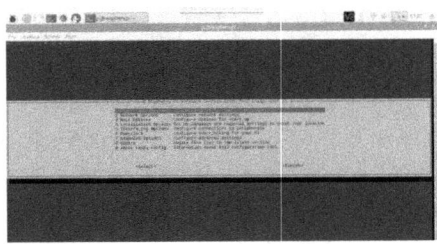

Interface of raspi-config on Raspbian terminal

raspi-config allows you to modify some features in order to simplify common Raspbian configurations. This tool automatically modifies the file contained in */boot/config.txt* and other standard Linux files that deal with operating system configuration.

Use of raspi-config

We use the up and down arrows to move within the raspi-config environment. A highlighted line will move between the various available options. With the left and right arrows, you can enter and exit each menu. Alternatively, to use the buttons that the environment provides, you can use the *TAB* button.

A numbered list shows a menu that allows you to activate or deactivate functions or features of the board. The first point, highlighted in the previous image, allows you to change the password of the raspbian user (which can still be done using the *sudo passwd pi* command). After selecting this first point, the environment will ask us to type the user's new password and then confirm it.

The second point of the raspi-config environment allows you to configure network-level settings on our board. Once point two is selected, a screen will be shown as follows:

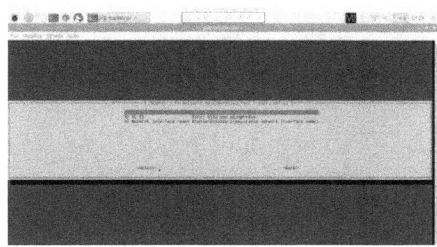

Network configuration via raspi-config

To change the name of our device in the local network, click on the hostname, highlighted in red in the previous figure.

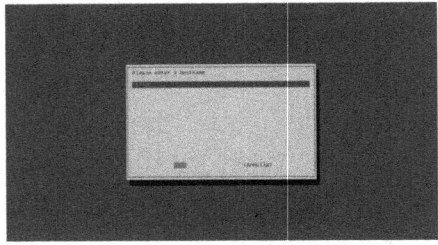

Inside the blue input we insert the name you want to assign to the board (in this case htmlpi has been inserted). Then move to the <OK> tab with the *TAB* key and press the Enter key. Still, inside the second point of raspi-config we can connect our device to the wireless network, obviously knowing the SSID, or more commonly the name of the wireless network, and the passphrase.

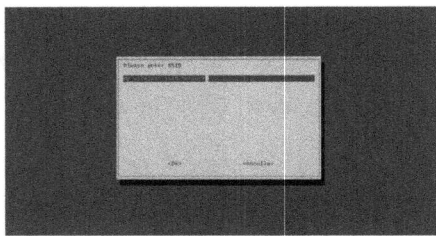

Entering the SSID to connect to the wireless network using raspi-config

So let's insert the name of our wireless network inside the blue input, and move to the <OK> tab with the *TAB* key and press the Enter key. At this point, the environment will ask us the passphrase of our wireless network.

Once the passphrase is inserted, move to the <OK> tab with the TAB key and press the Enter key. If everything went correctly, a screen will confirm the success of the operation and the successful connection.

Interface Configurations

In the previous chapter, we saw how to configure the interfaces via a graphical user interface. Below we will see how to do the same things with raspi-config. Select the *Interfacing Options* using the right arrow or selecting *<Select>* using the *TAB* key and pressing Enter. The tool will show an interface like this one:

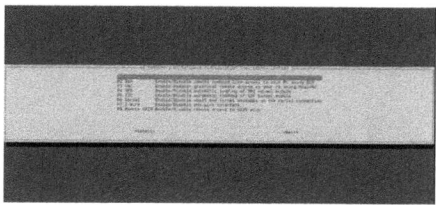

Interfacing Options in raspi-config on Raspbian

Each line allows you to enable or disable ports and services within the Raspberry Pi. In particular, to enable the Pi camera connected to the CSI connector of the board itself, select the corresponding line (as in the previous figure) and presses Enter.

We are asked if we want to enable the interface. We answer *<Yes>*.

A confirmation window will be shown to confirm the success of the operation. Like the camera, any other service or interface can be enabled/disabled following the same steps.

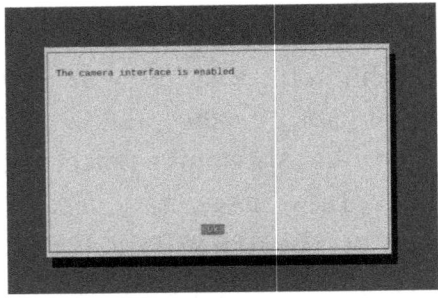

For further information about the raspi-config configuration tool, please refer to the appropriate page on the official website of Raspberry Pi.

Chapter 5

Headless Raspberry Pi Setup

This chapter describes how to configure the Raspberry Pi in headless mode, so it will allow you to remotely access it via SSH (Secure Shell) without the use of external monitors, keyboard, and mouse. It will describe the steps to follow and finally introduce a method to access the Raspberry Pi without knowing its IP address.

The Raspberry Pi is part of the Single Board Computer (SBC) category and it is, therefore, a full-fledged computer, very similar to the PC and Laptop that we all know.

Using the Raspberry Pi in headless mode may be necessary, or otherwise very beneficial, when you build, for example, a media server, web server, NAS, Samba server, or any network service that does not require direct user interaction with the Raspberry Pi. Therefore, in this mode, the configuration of the Raspberry Pi will be done through the network connection via the SSH protocol. You may be wondering how you can set up wireless network access to the Raspberry Pi without using the keyboard and monitor to perform network configuration in Raspbian windows. We will plot an effective path to achieve this.

Here are the simple steps you need to take to set up your wireless network on the Raspberry Pi:

1. **Download Raspberry Pi OS Buster lite.**

   You can download the image here: https://www.raspberrypi.org/downloads/raspberry-pi-os/

2. **Burn the Raspberry Pi OS image to the SD card.**

   Using the free **etcher** software, burn the previously downloaded image to a microSD card.

   Here the steps on how to use Etcher:
   1. Browse to https://www.balena.io/etcher/;
   2. Download the version for your operating system;
   3. Run the installer.
      Put a blank mini SD card and adapter into your machine. There's no need to format it. You can use a new SD card right out of the package.
      After you flash (burn) the image, Finder (Mac) or File Explorer (Windows) may have trouble seeing it. A simple fix is to pull the SD card out then plug it back in. On a Mac, it should appear on the desktop with the name boot. On Windows, it should appear in File Explorer with the name boot followed by a drive letter.

3. **Enable ssh to allow remote login.**

   For security reasons, ssh is no longer enabled by default. To enable it you need to place an empty file named ssh (no extension) in the root of the boot disk.

4. **Add your WiFi network info.**
   Create a file in the root of boot called: *wpa_supplicant.conf* (instructions below). Then paste the following into it (adjusting for your ISO 3166 alpha-2 country code, network name and network password):

   country=US
   ctrl_interface=DIR=/var/run/wpa_supplicant GROUP=netdev
   update_config=1
   network={
   ssid="NETWORK-NAME"
   psk="NETWORK-PASSWORD"
   }

Mac instructions:
Create a new empty file that will hold network info:

   touch /Volumes/boot/wpa_supplicant.conf

Edit the file that you just created and paste the text above into it (adjusting for the name of your country code, network name and network password).

Windows instructions:

1. Run Notepad;
2. Paste in the contents above (adjusting for the name of your country code, network name, and network password);
3. Click *File / Save As ...*;
4. Be sure to set Save as type to All Files (so the file is NOT saved with a .txt extension);

5. Call the file wpa_supplicant.conf and save it;
6. Close the file.
1. **Eject the micro SD card and boot the Raspberry Pi from the micro SD card.**
   Remove the microSD card from your computer and plug it into the Raspberry Pi then turn it on for the first time. For the Raspberry Pi 4, you need to plug a USB-C power supply cable into the power port.
   After a few seconds, if all went well, the Raspberry Pi will finish the boot phase and wait for an SSH connection from the outside. The last information needed to access the Raspberry is the IP address assigned to it by the Wi-Fi router

Login With IP Address

To find out the IP address, we can connect to the Wi-Fi router and check the assigned address on the appropriate page (each router displays this information on a different page). The default user is called pi, so you can remotely access the Raspberry Pi with the command:

ssh pi@<IP address>

Alternatively, you can use any SSH client. For example on Windows the SSH Putty client.

Login with Hostname

Please note that some routers register the hostnames of devices that have made a DHCP request in their local DNS. Since Raspberry Pi by default uses the name **raspberrypi** as the hostname, you can access Raspberry Pi via SSH without knowing its IP address, but only using the hostname.

In this case, you access the Raspberry Pi via SSH with:

ssh pi@raspberrypi

Login With mDNS Name

In case your Wi-Fi router does not register hostname in the local DNS, you can take advantage of a convenient network service called mDNS (multicast DNS). Raspbian provides a very useful feature that allows access to the Raspberry Pi using the name raspberrypi.local instead of the IP address. It is important to know that on your computer will have to be active the mDNS service typically provided by Apple Bonjour that is installed with iTunes.

If you do not want to install iTunes, you can only use the Bonjour service. Download Bonjour Print Services for Windows on *support.apple.com* and simply run the installer.

In this case, you can access the Raspberry Pi via SSH with:

ssh pi@raspberrypi.local

In conclusion, by following these instructions you can configure the Raspberry Pi in the headless mode which will allow us to immediately exploit the full potential of the device without passing through input/output devices physically connected to it.

Chapter 6

Programming the Raspberry Pi

In previous chapters, we have seen how to install Raspbian on our Raspberry Pi. This operating system, in addition to managing Raspberry and the main peripherals, allows you to program the board in its entirety. Among other things, in fact, it allows you to manage the GPIO port on either side of the Raspberry Pi. This is possible through different programming languages, including **Python**, Java, C, and C++. In the next chapters, we will use the Python language for the development of the projects we will discuss in the guide. In addition to being one of the most widely used languages (especially recently), Python also has the advantage of having a strong community of developers, which helps to release many freely accessible libraries. Among them, **RPi.GPIO** is the one that allows you to interact with the GPIO port of Raspberry, and is, therefore, the main programming tool.

In this chapter, we will first introduce the different development environments and then continue by referring to the Python programming language.

The operating system installed on our board allows us to interact with the outside world through peripherals that can be connected to the ports of our mini-computer, or with electrical circuits made by us through the use of sensors and electronic components that you can buy anywhere (even at low prices). Programming a device like Raspberry Pi is easy even for those who have never had any experience of programming on personal computers or using operating systems from the Linux world.

The Raspberry Pi Foundation, which aims to help as many people as possible approach the world of programming, has equipped the Raspbian operating system with many simple, powerful and intuitive development environments that accompany the user during this training. Among them, we have **BlueJ** and **Greenfoot** for development using the Java language, and **Scratch**, a graphical development environment suitable for those who are taking their first steps into the world of programming.

Development

You can generally follow three different modes to run the script on Raspberry:

1. **Direct execution on the Raspberry Pi.**
   In this case, the board is connected to the screen, mouse, keyboard, and any other necessary peripheral. Having full control, you can proceed with the same programming modes used on a standard installation of a Debian-based operating system.

2. **Secure Shell connection (SSH).**
   To use this type of connection, it is strictly necessary to have connected the Raspberry to our home network

(through the Wireless module, or via Ethernet cable). Then, you must start the SSH server on Raspberry and connect using any SSH client (Putty on Windows, or a command-line solution on Linux and macOS systems). In the case of Linux and MacOS, the SSH connection can be made by typing a string like the following:

ssh user@ip-address-raspberry

3. **VNC connection**.
Also to use the VNC connection, you must first connect the board to your home network. In recent versions of Raspbian, there is an integrated VNC server that can be enabled during installation or through raspi-config: to use the latter tool, open the terminal and type:

raspi-config

Next, follow the path *raspi-config > interfacing options > VNC > YES*. The VNC server provides the Raspbian GUI through a VNC client installed on another computer, from which we can use Raspberry and then program it.

In all three cases, we will have access to the terminal, and therefore to the shell. So let's make sure Python is correctly installed by typing *python* and pressing Enter. If everything is configured correctly, the Python interpreter will be loaded, showing the version and the characters >>>.

In the following, we will not dwell on the use of Python, but we will assume that the syntactic basis of the language is clear. In any case, to develop scripts in Python you just need a simple text editor, but there are also more sophisticated development environments, such as **PyCharm** or **PyDev**. Once our script is created, we will be able to run it from theo terminal (or through one of the above-mentioned environments).

Python Modules

Before continuing, we will have to install the modules necessary for the operation of a script that can manage GPIO. To do this, we can use **pip**, a package installation tool for Python. The installation is very simple, just type from the terminal:

wget https://bootstrap.pypa.io/get-pip.py
sudo python get-pip.py

When the tool has been installed you can install the packages in the repository through the command:

pip install package-name

Let's install RPi.GPIO as follows:

sudo pip install RPi.GPIO

This library may already be inside Python for Raspberry, but since it is vital to our work, if it is not already installed, the command will proceed with the installation.

Chapter 7

LCD Alphanumeric Display

In this chapter, we will learn how to control an alphanumeric LCD using the Raspberry Pi. We will analyze all the necessary steps to make the electrical connections, we will also describe how to install the python library suitable for controlling the display. Then we will write a "base" code to manage it.

On the market, there are various types of displays that can be used with the Raspberry Pi, ranging from the simplest alphanumeric LCD (Liquid Crystal Display) to the most sophisticated touch screens.

Let's see the main types of displays that can be used:

1. **Alphanumeric LCD**.
   Normally organized in lines of characters. Available in various sizes including the most common 2 or 4 lines of 16 or 20 characters (16×2 - 16×4 or 20×2 - 20×4). They display ASCII characters and allow you to have a customizable character map. They are easy to find, have a very low cost, and are easy to program.

2. **Graphical LCD.**

   Similar to the previous ones but with the fundamental difference that they do not display characters but single pixels. They exist in various sizes, among which the most common are 128×64 pixels, 122×32 pixels, or 84×48 pixels. They allow you to display simple images or graphic effects. Also, this type of display is inexpensive and widespread in the maker field.

3. **OLED (Organic Light Emitting Diode).**

   Small graphic displays both monochrome and color with OLED technology. Similar to the previous ones as functionality, but normally smaller and thinner.

4. **TFT (Thin Film Transistor).**

   They are active-matrix displays used in past years for laptop screens. Normally of larger size (3 to 7 inches), in color, they can also be touch-screen. Generally, they have a higher cost than the previous ones and are more suitable for applications where there is a strong graphic component.

5. **E-Ink.**

   Display commonly used by e-book readers. A key feature is the low power consumption and the ability to display static content even in the absence of power. Typically they have a higher cost and are used in particular applications (often outside the maker field).

6. **Led matrix**.

   Monochromatic or RGB LED matrix. It exists in various sizes and the most used in the maker are modules of 8×8 pixels modular. Typically used for the display of scrolling messages and writing. It has a very low cost and is easy to find and use. The most common displays can be controlled via the Raspberry Pi using simple parallel protocols: *I2C (Inter-IC)* or in some cases via *SPI (Serial Peripheral Interface)* connections. More advanced displays, such as *TFT*, have HDMI connections.

Control of an Alphanumeric LCD

The most popular alphanumeric LCDs in the maker industry are those based on the *Hitachi HD44780* driver. Therefore in this lesson, we refer to this type of display. These LCDs are supplied by a wide range of suppliers, including all those present on various e-commerce sites for electronic components.

For this reason, although they are all functionally the same and based on the same control driver, they may present minimal differences. The most important difference could be the arrangement of the control pins. Let's now describe all 16 pins with name and functionality:

Pin #1[Vss]	Bulk
Pin #2[Vdd]	Power supply
Pin #3[Vo]	LCD contrast control
Pin #4[RS]	Register Select
Pin #5[R/W]	Read/Write mode
Pin #6[E]	Enable
Pin #7[DB0]	Bit 0

Pin #8[DB1]	Bit 1
Pin #9[DB2]	Bit 2
Pin #10[DB3]	Bit 3
Pin #11[DB4]	Bit 4
Pin #12[DB5]	Bit 5
Pin #13[DB6]	Bit 6
Pin #14[DB7]	Bit 7
Pin BL-	back-lighting
Pin BL+	+5V back-lighting

Looking at the LCD display from above, with the connector facing down, starting from the right of the connector, the pins are arranged in this order: BL+, BL-, #1, #2, #3, #4, #5, #6, #7, #8, #9, #10, #11, #12, #13, #14. There are some variations in the arrangement of these pins but following the name and numbering the functionality of each pin remains unchanged. To realize the control circuit, in addition to the LCD and the Raspebrry Pi, a 10K Ohm potentiometer, a 220 Ohm resistor, and 20 multi-colored male-female wires are required. The circuit and the connections to be made are shown in the figure.

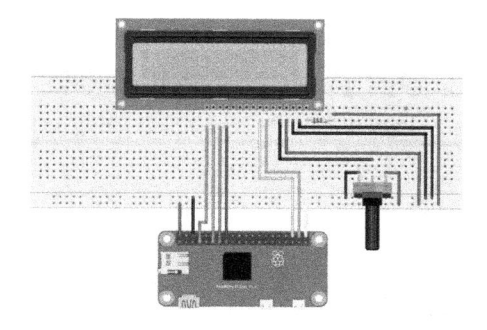

To optimally connect the Raspberry Pi to the LCD, make the following connections using physical pin numbering:

Raspberry pin #2	line + red on breadboard
Raspberry pin #6	line - blue on breadboard
Raspberry pin #7	LCD pin #14 [DB7]
Raspberry pin #11	LCD pin #13 [DB6]
Raspberry pin #13	LCD pin #12 [DB5]
Raspberry pin #15	LCD pin #11 [DB4]
Raspberry pin #36	LCD pin #6 [E]
Raspberry pin #38	LCD pin #5 [R/W]
Raspberry pin #40	LCD pin #4 [RS]
LCD pin#3 [Vo]	center potentiometer pin
LCD pin#2 [Vdd]	line + red on breadboard
LCD pin#1 [Vss]	line - blue on breadboard
LCD pin# BL-	line - blue on breadboard
LCD pin# BL+	220 Omh resistor
Other resistor	line + red on breadboard
Right pin potentiometer	line + red on breadboard
Left pin potentiometer	line - blue on breadboard

The potentiometer is used to adjust the contrast of the LCD. It is suggested to use it in case the characters are not fully visible.

Control Software

To control the LCD, with the Raspberry Pi, we use a python library called RPLCD. The library supports different types of displays and therefore it must be configured correctly according to the display used in your project. You can install the library from a terminal window with the following command:

sudo pip3 install RPLCD

The use of python3 is assumed, however, the library is also compatible with python2. Below is the commented code of the control program.

```python
import sys
from RPLCD.gpio import CharLCD
from RPi import GPIO
import time
DELAY = 0.5
lcd = CharLCD(numbering_mode=GPIO.BOARD, cols=16, rows=2, pin_rs=40, pin_rw=38, pin_e=36, pins_data=[15, 13, 11, 7])
#set back-lighting / hide pointer
lcd.backlight = True
lcd.cursor_mode = 'hide'
lcd.clear(). #clear display
lcd.cursor_pos = (0, 5). #move pointer row 1, column 5
lcd.write_string("Hello") #print string
lcd.cursor_pos = (1, 1) #move pointer row 2, column 1
lcd.write_string("Raspberry Pi!")
time.sleep(1)
#Scroll left and right writing
for _ in range(16):
lcd.shift_display(-1)
time.sleep(DELAY)
while True:
for _ in range(32):
lcd.shift_display(1)
time.sleep(DELAY)
for _ in range(32):
lcd.shift_display(-1)
time.sleep(DELAY)
```

The **CharLCD** class provides a simple interface to the LCD display and allows you to control every single aspect. As for the electrical connections, the code uses the physical numbering for the pins of the Raspberry Pi. In the class *constructor* you have to pass some basic parameters:

1. **cols and rows**.
   They are the number of columns and rows of your LCD.
2. **pin_rs**.
   It corresponds to LCD pin #4 and it is connected to Raspberry pin#40.
3. **pin_rw**.
   It corresponds to LCD pin #5 and it is connected to Raspberry pin#38.
4. **pin_e**.
   It corresponds to LCD pin #6 and it is connected to Raspberry pin#36.
5. **pin_data**.
   A list of 4 or 8 pins corresponding to the DB0-7 pins of the LCD. In this case, a 4-bit data mode is used to reduce the number of connections.

If all the steps have been performed correctly, you should get a result similar to the figure below.

In this chapter, we have done an overview of the most common types of displays for Raspberry Pi in the maker industry. Then we described in detail the pinout of the LCD, with *Hitachi HD44780 driver*, and described the electrical connections necessary for its proper operation. A control and verification code, of the operation of the LCD, was provided and analyzed just above, the latter can be used as a basis for your implementations.

Chapter 8

DHT11 Humidity and Temperature Sensor

In this chapter, we will learn how to use the DHT11 digital sensor to detect ambient temperature and atmospheric pressure via the Raspberry Pi. We will see what electrical connections are needed and use a software control to read the environmental values.

DHT11 sensor

The *DHT11* component is a temperature and humidity sensor with digital output that implements a 1-wire protocol. In other words, it allows communicating the detected values to a Raspberry Pi up to 20mt distance using a single wire in addition to the power supply.

This feature is very important because the Raspberry Pi is not equipped with analog pins and therefore it is not immediately possible to use such sensors. The sensor is specially calibrated by the manufacturer in order to provide stable measurements. The DHT11 operates in the temperature range of 0 to 50 degrees Celsius and in the range of 20% to 90% relative humidity. It is commercially available with 3 or 4 functionally equivalent pins. In the 4-pin version, one of them is not connected. The DHT11 must be powered with a voltage between 3 and 5V. To use the DHT11 sensor with the Raspberry Pi we use the *Adafruit_DHT* library. To install it just run the following command:

sudo pip3 install Adafruit_DHT

DHT11 Control Circuit

The *DHT11* sensor exposes 3 pins named:

1. GND – ground pin;
2. VCC – power supply pin;
3. DAT - 1-wire data communication pin.

To realize the control circuit, in addition to the Raspberry Pi, a 4.7 KOhm resistor, the DHT11 sensor, and 3 male-female rainbow wires are required. The circuit and the connections to be made are shown in the figure.

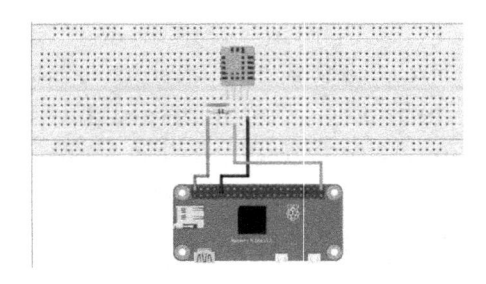

Connections to be made:

Raspberry pin #2	VCC pin
Raspberry pin #9	GND pin
Raspberry pin #40	DAT pin
Raspberry pin #11	LCD pin #13 [DB6]
Between VCC and DAT pin of DHT11	Insert the 4.7 KOhm resistor to stabilize the data line

GPIO PIN Numbering

Pin numbering, in the various versions of Raspberry Pi that have followed one another over the years, has always been a controversial issue. The main reason is that different libraries adopt different numbering. Also, different versions of the Raspberry Pi have different GPIO connectors and numbering.

Specifically, there are two types of GPIO pin numbering:

1. **Physical numbering.**
   Pins are numbered sequentially by column, assigning number 1 to the first pin at the bottom left and number 2 to the first pin at the top left (looking at the connector from above). In the figure, the numbers inside the colored circle represent the physical numbering of the GPIO pins.
2. **BCM numbering.**
   The pin numbering follows the one defined by the Broadcom chip of the Raspberry Pi. BCM in fact means Broadcom SOC channel. In the figure, the BCM numbering is shown in the rectangle below each pin and preceded by the GPIO prefix.

For example, pin 7 in physical numbering corresponds to GPIO pin 4 in BCM numbering. Physical pin number 40 corresponds to GPIO 21 in BCM numbering.

It is important to note that the numbering in the figure applies only to the following models:

- Raspberry Pi Model B+;
- Raspberry Pi 2B;
- Raspberry Pi Zero;
- Raspberry Pi Zero W;
- Raspberry Pi 3B;
- Raspberry Pi 4.

Some libraries allow you to set which numbering to use.

Temperature and Humidity Reading Code in Python

Below is the python code to run on the Raspberry Pi to perform ambient temperature and humidity measurement.

```
import Adafruit_DHT # 1
import time # 2

DHT11 = Adafruit_DHT.DHT11 # 4
DHT11_PIN = 21 # BCM pin numbering # 5

while True: # 7
humidity, temperature = Adafruit_DHT.read_retry(DHT11, DHT11_PIN, retries=2, delay_seconds=1) # 8
if humidity is not None and temperatura is not None: # 9
print('Temperature={0:0.1f}*(     Humidity={1:0.1f}%'.format(temperature, humidity)) # 10
else: # 11
print(Error in acquiring values. Try again!') # 12
time.sleep(5)
```

Let's break down the code:

1. **Line 4** - the variable DHT11 is set with the specific type of sensor used. This is necessary because the library we use supports different types of sensors.
2. **Line 5** - the variable DHT11_PIN contains the interface pin, in BCM numbering, between the sensor and the Raspberry Pi.
3. **Line 8** - through the read_retry method we read the values detected by the sensor. Please, note that this method has two very important parameters:
   1. *retries* - number of attempts in case of a failed read. The 1-wire protocol implemented requires the respect of very stringent timing. Linux, not being a hard real-time operating system can, at times, not respect the parameters and therefore not perform the reading correctly.

2. *delay_seconds* - waiting time between readings.
4. **line 9-12** - According to what has been explained the values of humidity and temperature can be null. So the code prints the parameters only if they are actually detected. Otherwise, it displays an error message.
5. **line 13** - waits 5 seconds before performing a new reading.

In this chapter, we saw how easy it is to interface with the DHT11 digital temperature and humidity sensor. We have analyzed the few necessary electrical connections and the control code in python. In the next chapter, using what we learned in the previous chapter, we will build a simple weather station with LCD that shows temperature and humidity measured near the Raspberry Pi. Finally, through the *Yahoo Weather* cloud service, we will extend the functionality of the weather station by displaying the atmospheric parameters measured in a remote location.

Chapter 9

Mini Weather Station - Part 1

This chapter describes the implementation of a **mini weather station with the Raspberry Pi and the DHT11** sensor that displays locally measured temperature and humidity. In the next chapter, we will extend the functionality of the mini weather station by adding temperature and humidity detection at a remote location chosen by the user. Both chapters build on what was described in the previous chapters: DHT11 humidity and temperature sensor and alphanumeric LCD.

Mini Weather Station: Let's Build It!

To realize the weather station we must integrate on a single breadboard the circuit to control the DHT11 sensor with the one to control the LCD display. In this way, we will read the atmospheric parameters through the DHT11 sensor and display them on the LCD. The circuit and the connections to be made are shown in the figure.

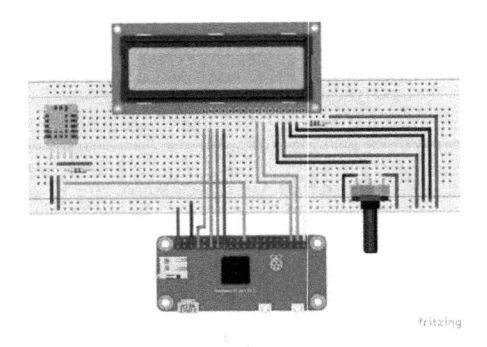

The connections to be made are as follows:

Raspberry pin #2	line + red on breadboard
Raspberry pin #6	line - blue on breadboard
VCC track on the breadboard	VCC pin
GND track on the breadboard	GND pin
Raspberry pin #22	DAT pin
Between VCC and DAT pin of DHT11	insert the 4.7 KOhm resistor to stabilize the data line
Raspberry pin #7	LCD pin #14 [DB7]
Raspberry pin #11	LCD pin #13 [DB6]
Raspberry pin #13	LCD pin #12 [DB5]
Raspberry pin #15	LCD pin #11 [DB4]
Raspberry pin #36	LCD pin #6 [E]
Raspberry pin #38	LCD pin #5 [R/W]
Raspberry pin #40	LCD pin #4 [RS]
LCD pin#3 [Vo]	center potentiometer pin
LCD pin#2 [Vdd]	line + red on breadboard
LCD pin#1 [Vss]	line - blue on breadboard
LCD pin# BL -	line - blue on breadboard

LCD pin# BL +	220 Omh resistor
Other resistor	line + red on breadboard
Right pin potentiometer	line + red on breadboard
Left pin potentiometer	line - blue on breadboard

Control Code in Python

Let's now analyze the weather station code, which is slightly more complex than in previous chapters.

```
import sys
import time
import Adafruit_DHT
from RPLCD.gpio import CharLCD
from RPi import GPIO
DELAY = 60
# DHT11 setup
DHT11 = Adafruit_DHT.DHT11
DHT11_PIN = 25
DHT11_ATTEMPTS = 2
DHT11_DELAY_ATTEMPTS = 1
def init_lcd():
    #Init display LCD
    lcd = CharLCD(numbering_mode=GPIO.BOARD, cols=16, rows=2, pin_rs=40, \
    pin_rw=38, pin_e=36, pins_data=[15, 13, 11, 7])
    lcd.backlight = True
    lcd.cursor_mode = 'hide'
    return lcd
def create_icons(lcd):
    #declaring icons
    icon_in = (
    0b01000,
    0b01100,
    0b01110,
    0b01111,
    0b01111,
    0b01110,
    0b01100,
    0b01000
    )
    icon_out = (
    0b00010,
```

```
    0b00110,
    0b01110,
    0b11110,
    0b11110,
    0b01110,
    0b00110,
    0b00010
)

lcd.create_char(0, icon_in)
lcd.create_char(1, icon_out)
def splash_screen(lcd):  #splash screen view
    lcd.clear()
    lcd.cursor_pos = (0, 3)
    lcd.write_string("Raspberry")
    lcd.cursor_pos = (1, 1)
    lcd.write_string("Weather Station")
    time.sleep(2)
    for _ in range(16):
        lcd.shift_display(-1)
        time.sleep(0.2)
    lcd.clear()
def show_values(lcd, humidity, temperature):
    lcd.cursor_pos = (0, 0)
    lcd.write_string("\x00T:{}C\x01".format(int(temperature)))
    lcd.cursor_pos = (1,0)
    lcd.write_string("\x00H:{}%\x01".format(int(humidity)))
def show_error(lcd):
    lcd.cursor_pos = (0, 5)
    lcd.write_string("Error")
    lcd.cursor_pos = (1,1)
    lcd.write_string("Read Values")
def lettura_DHT11():
```

```
return Adafruit_DHT.read_retry(DHT11, DHT11_PIN, \
retries=DHT11_ATTEMPTS, \
delay_seconds=\
DHT11_DELAY_ATTEMPTS)
def main():
lcd = init_lcd()
create_icons(lcd)
splash_screen(lcd)
while True:
humidity, temperature = read_DHT11()
if humidity is not None and temperature is not None:
print('Temperature={0:0.1f}*C Humidity={1:0.1f}%'\
.format(temperature, humidity))
show_values(lcd, humidity, temperature)
else:
print('Error getting values. Try again!')
show_error(lcd)
time.sleep(DELAY)
if __name__ == "__main__":
main()
```

Some control variables are defined at the top of the code:

1. **DELAY**.
   It defines the interval between two local parameter readings.
2. **DHT11**.
   It defines the type of sensor to correctly configure the library.
3. **DHT11_PIN**.
   1-wire data pin used to communicate with the DHT11 sensor.

4. **DHT11_ATTEMPTS.**
   The maximum number of attempts to read the environmental parameters before generating an error.
5. **DHT11_DELAY_ATTEMPTS.**
   Time in seconds between one reading attempt and the next.

main() method - As soon as the program runs, the main method is executed, which implements the main cycle of the weather station. The LCD is initialized, exactly as we have seen in the dedicated chapter. Next, custom icons are defined in the *create_icons()* method. At this point, the *splash_screen()* method displays an initial screen on the LCD to welcome the user. Within the main loop, via the *read_DHT11()* method, values are initially read from the sensor. If the values are valid, then they have displayed on the LCD via the *display_values()* method. Otherwise, the display_error method is invoked so that an error message appears on the LCD.

init_lcd() method - The constructor of the Adafruit_DHT library is invoked. The meaning of the individual parameters was described in the Alphanumeric LCD chapter so it is not repeated here.

create_icons() method - The LCD display and control library allows a user-defined character map to be set. In this case, we define two icons that represent whether the atmospheric values were collected locally or remotely. *icon_in* and *icon_out* are two tuples of binary values that represent the icon bitmap. The *create_char()* method takes care of the actual creation of the custom characters in the memory map of the LCD.

splash_screen() method – This method, invoked only during the initialization of the program, displays an initial screen and, through the *shift_display()* method, realizes an animation/scroll effect of the text on the display.

show_values() method – This method displays the temperature and humidity values. The *write_string()* method is used to display a string on the display. Note the way the icons we defined earlier are inserted in the string. In particular *\x00* refers to the first icon defined, while *\x01* indicates the icon in the second memory zone of the user character map.

show_error() method – In case of an incorrect reading by the sensor, this method is invoked to inform the user of what happened.

read_DHT11() method - The *read_retry()* method of the *Adafruit_DHT* library reliably reads atmospheric parameters via the DHT11 sensor.

In conclusion, following this chapter, with a few simple components, you can build a mini weather station that allows you to measure the temperature and humidity in the vicinity of the Raspberry Pi. In the next chapter, using the *Yahoo Weather* library, we will extend the functionality of the weather station by adding the ability to measure temperature and humidity in a remote location defined by the user.

Chapter 10

Mini weather Station - Part 2

In this second part, we describe how to extend the functionality of the mini weather station by adding temperature and humidity detection at a user-defined remote location.

Detection of Remote Environmental Values

In order to detect temperature and humidity at a remote location, we rely on the *Yahoo Weather* library.

To install it just run the following command:

sudo pip3 install weather-api

Let's now analyze the changes to be made to the weather station code in order to make the reading through the Yahoo Weather library.

```python
import sys
import time
import Adafruit_DHT
from RPLCD.gpio import CharLCD
from RPi import GPIO
from weather import Weather, Unit
DELAY = 5
LOCATION = "London"
# DHT11 setup
DHT11 = Adafruit_DHT.DHT11
DHT11_PIN = 25
DHT11_ATTEMPTS = 2
DHT11_ATTEMPTS_DELAY = 1
def init_lcd():
    #Init display LCD
    lcd = CharLCD(numbering_mode=GPIO.BOARD, cols=16, rows=2, pin_rs=40,\
    pin_rw=38, pin_e=36, pins_data=[15, 13, 11, 7])
    lcd.backlight = True
    lcd.cursor_mode = 'hide'
    return lcd
def create_icons(lcd):
    #declaring icons
    icon_in = (
    0b01000,
    0b01100,
    0b01110,
    0b01111,
    0b01111,
    0b01110,
    0b01100,
    0b01000
    )
```

```
icon_out = (
0b00010,
0b00110,
0b01110,
0b11110,
0b11110,
0b01110,
0b00110,
0b00010
)

lcd.create_char(0, icon_in)
lcd.create_char(1, icon_out)
def splash_screen(lcd):  #splash screen view
lcd.clear()
lcd.cursor_pos = (0, 3)
lcd.write_string("Raspberry")
lcd.cursor_pos = (1, 1)
lcd.write_string("Weather Station")
time.sleep(2)
for _ in range(16):
lcd.shift_display(-1)
time.sleep(0.2)
lcd.clear()
def show_values(lcd, local_humidity, local_temperature,\
weather_humidity, weather_temperature):
lcd.cursor_pos = (0, 0)
lcd.write_string("\x00T:{}C\x01 \x01T:{}C\x00"\
.format(int(local_temperature), int(weather_temperature)))
lcd.cursor_pos = (1,0)
lcd.write_string("\x00H:{}%\x01 \x01H:{}%\x00"\
.format(int(local_humidity),\
int(weather_humidity)))
```

```
def show_error(lcd):
lcd.clear()
lcd.cursor_pos = (0, 5)
lcd.write_string("Error")
lcd.cursor_pos = (1,1)
lcd.write_string("Read values")
def read_DHT11():
return Adafruit_DHT.read_retry(DHT11, DHT11_PIN,\
retries=DHT11_ATTEMPTS,\
delay_seconds=\
DHT11_ATTEMPTS_DELAY)
def init_weather_api ():
meteo = Weather(unit=Unit.CELSIUS)
return meteo
def read_weather (weather, location):
lookup = weather.lookup_by_location(location)
temperature = lookup.condition.temp
humidity = lookup.atmosphere.humidity
return humidity, temperature
def main():
lcd = init_lcd()
create_icons(lcd)
splash_screen(lcd)
meteo = init_weather_api ()
while True:
local_humidity, local_temperature = read_DHT11()
if local_humidity is not None and \
local_temperature is not None:
weather_humidity,    weather_temperature    =    read_weather(weather,\
LOCATION)
print('Local Temperature={}*C Local Humidity ={}%'\
.format(int(local_temperature), \
```

```
       int(local_humidity))))
       print('Weather Temperature={}*C Weather Humidity ={}%'\
       .format(weather_temperature, weather_humidity))
       show_values(lcd, local_humidity, local_temperature,\
       weather_humidity, weather_temperature)
       else:
       print('Error getting values. Try again!')
       show_error(lcd)
       time.sleep(DELAY)
       if __name__ == "__main__":
       main()
```

Let's now describe the changes made to the code from the previous version. We have added a single control variable.

1. **LOCATION**.
It defines the remote location from which to collect data.

main() method - The only change made is to read humidity and temperature in the place defined by the method *read_weather()*.

show_values() method – Conceptually the same as before but now displays 4 values instead of 2.

init_api_weather() method - This method calls the constructor of the Weather library. The unit of measurement (degrees Celsius) is passed as an initialization parameter.

read_weather() method - Through the *lookup_by_location()* method the actual API call for detecting remote values is made. Finally, temperature and humidity are extracted from the object returned by the call.

Let's interpret values. As you can now see on the display, 4 values are shown. Two of temperature (T) and two of humidity (H). The values on the left side of the display, with the converging icons, represent the local values. Those on the right are the values taken via API at the chosen remote location.

In this chapter, we completed the implementation of the mini weather station using the concepts and knowledge learned in the previous chapters. This allowed us to integrate the DHT11 sensor and LCD into one circuit and design. After that, we extended the weather station code, via a python library, thus adding the ability to display both local and remote location environmental data.

Chapter 11

GPIO Connector Installation

This chapter describes how to physically access the GPIO pins on the Raspberry Pi Zero, that is, **how to install and solder the GPIO connector** to the board. It explains the detailed steps to follow and provides some useful tips to complete the job effectively and quickly.

GPIO Interface

The Raspberry Pi is the most well-known and popular single-board computer developed in the UK by the RaspberryPi Foundation. The first version called Raspberry Pi 1 dates back to 2012. Since then, different versions and variants of the board have been developed, with different features in terms of CPU, memory, connectors, and size. A key feature, which unites all versions of the Raspberry Pi board, is the presence of a *General Purpose Input/Output (GPIO) connector*. To be clear, even the very popular microcontroller board Arduino, has a GPIO connector.

The GPIO interface allows the inter connection of the microprocessor, or microcontroller in the case of Arduino, with one or more external digital devices. As the name suggests, through the GPIO interface, you can read (INPUT) the status of an external device, or set a value (OUTPUT) always on an external device. For example, you can read the status of a button, connected to a GPIO pin of the Raspberry Pi, and then perform a specific action. In the output, always through a GPIO pin, you can set the status of a LED by turning it on and off.

Other examples of INPUT are:

- External temperature reading via a digital thermometer;
- External barometric pressure reading;
- Reading a serial message sent by another device;
- Reading a switch or breaker.

Examples of OUTPUT are:

- Switching on a traditional LED;
- WS2812 Digital Led Control;
- Writing on an alphanumeric display;
- Drawing on a graphic display;
- Speed control of a DC motor by means of a special driver.

All Raspberry Pi models are sold with the GPIO connector already installed. Exceptions are the two smaller boards Raspberry Pi Zero and Raspberry Pi Zero W.

Raspberry Pi Zero

Let me digress a bit about the Raspberry Pi Zero and the connector installation. You can skip this paragraph if you own a different model of Raspberry. Normally the Raspberry PI Zero (Zero and Zero W models) is sold without the 40-pin GPIO connector, which is an optional component that can be purchased separately. The lack of the GPIO connector is not a problem in itself and does not affect in any way the operation of the Raspberry Pi. Clearly, it is impractical, if not impossible, to use the Raspberry Pi in projects in which you make use of GPIO pins, for example, to control an external relay.

The board has two rows of 20 holes each on the top edge. These are the pads on which it is possible to mount and solder a 40 pin connector, which is very easy to find commercially.

This connector is generically called 2 Row Pin Header (Male). The pitch of this connector is standard and measures exactly 2.54 mm. To be used and mounted properly on the Raspberry Pi Zero must have 20 pins per row and then 40 pins total.

RASPBERRY PI ZERO: INSTALLING THE 40-PIN GPIO CONNECTOR

The following describes how to install and solder the 40-pin GPIO connector on the Raspberry Pi Zero in a quick and convenient way.

To complete the installation of the GPIO connector, you will need the following components:

- Il Raspberry Pi Zero;
- The 40-pin male GPIO connector;
- Breadboard.

In addition, the following tools are required:

- A tin soldering iron with fine tip;
- A tin spool (Sn60/Pb40 - Gauge 22 or 0.7 mm).

Tin Soldering

Tin soldering of printed circuit boards is a fairly simple technique so it should not scare you. If you have a fine-tipped soldering iron and a tin spool, soldering the GPIO connector is fairly simple. It can be done by anyone with a minimum of practice.

In short, the steps to make a good weld are as follows:

1. Heat the soldering iron well, bringing the tip to a temperature of about 300/350° Celsius.

2. Place the tip of the soldering iron in contact with the pin and pad for about 1 second.

3. Melt a small amount of tin (about 1-2 mm in length) by placing the wire under the tip of the soldering iron.

4. As soon as the tin has melted, remove the tin wire to prevent excessive deposit.

5. Hold the soldering iron tip in place for another second so that the tin is evenly distributed over the pin and pad.

6. Remove the soldering iron tip, allowing the tin to solidify.

7. Make sure the solder is uniform and connects the pin to the pad well.

If the soldering is not satisfactory, you can remove the deposited tin with a small suction pump and solder again. We will now describe in detail the steps to follow to solder the GPIO connector to the Raspberry Pi Zero.

1. The first thing to do is to make sure that the Raspberry Pi's pads and GPIO connector pins are clean. If not, use a small piece of cotton or paper to clean them thoroughly.

2. Insert the 40-pin GPIO connector into the center of the breadboard taking care that all pins are properly inserted into the holes in the breadboard.
If the connector has been inserted correctly, it should be quite difficult to remove it by hand alone. If you do not have a breadboard available, you can use a "third hand."

However, this would make the soldering process more difficult so it is not recommended. The use of the breadboard allows you to create a stable base and to hold the 40-pin connector, thus facilitating the soldering process and the final quality of the work.

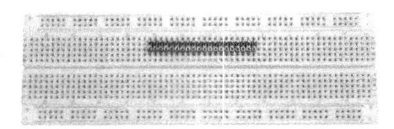

3. Turn the Raspberry Pi Zero upside down and insert the 40 holes so that they match exactly with the GPIO connector previously fixed on the breadboard. In this way, the Raspberry Pi is placed in a stable position on the breadboard so you can get more precise soldering. This little trick allows you to concentrate on soldering without having to worry too much about keeping the connector and the Raspberry Pi in a stable position.

4. After heating the soldering iron to a temperature of about 300/350° Celsius, start soldering the two pins in the opposite corners, as shown in the figure. This allows you to secure the Raspberry Pi PCB to the GPIO connector increasing the stability of the board.

It is very important to use a soldering iron that is well heated and has a properly sized tip. The space between the pins is small, but still sufficient to allow you to bring the tip of the soldering iron close to only one pin at a time.

5. Continue soldering all pins in an orderly fashion. Solder all the pins in the first row first and then move on to the pins in the second row.

As you can see in the figure, to make correct soldering, bring the tip of the soldering iron close so that it touches a pin and the pad. Then hold the soldering iron in contact with the pin for 1-2 seconds to properly heat it. Now bring the tin wire into contact with the pin and soldering iron tip so that a small drop of tin melts. Remove the tin wire and, after about 1 second, gently lift the soldering iron. The drop of tin at this point adheres to the PCB pad and solidifies quickly, thus soldering the connector pin to the Raspberry Pi. Once the soldering of all pins is completed, you should obtain a result similar to the one in the figure.

6. Pull the Raspberry Pi off the breadboard gently and calmly, avoiding bending one or more pins of the GPIO connector.

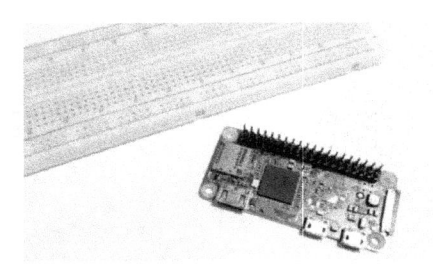

In this chapter, we have detailed the steps to be able to solder the 40-pin GPIO connector on the Raspberry Pi Zero board. We have also listed the components and tools to complete the task. At this point the Rapberry Pi Zero is ready, it can be used for an infinite number of projects that, through the GPIO pins use and drive external components. The use of GPIOs will be the subject of future chapters in this guide.

Chapter 12

GPIOZero Python Library

This chapter introduces the use of the GPIOZero python library to be able to program and use the GPIO pins of the Raspberry Pi. An overview of the main functionality of the library is given and the steps to follow for installation are described. In addition, two different GPIO pin numbering modes are described.

GPIO and Control Code

As described in the previous chapter of this guide, one of the main features of the Raspberry Pi is the ability to interface with external devices via the 40-pin GPIO connector. For example, it is possible to read (INPUT) from external sensors values such as temperature, humidity, and atmospheric pressure.

Through the GPIO it is also possible to control (OUTPUT) external devices, for example, to trigger a relay or drive an RGB led strip. Obviously, in addition to making the electrical connections, you must write and run a program to control the external devices and sensors. The Raspberry Pi is a full-featured computer that supports various operating systems, including Linux, so it is possible to write the control program in a multitude of programming languages. Younger people could use the visual language **Scratch** (highly recommended to introduce young people to the world of computing and programming). Those who are looking for maximum performance can program in **C/C++**. Those who come from the web world can use **node.js**. Those interested in using a booming language can write and control GPIOs in **go-lang**. Finally, for those looking for simplicity, expressiveness, and immediacy, the use of the **Python** language is recommended.

The GPIOZero Library

The GPIOZero python library for Raspberry Pi provides a simple interface to devices and sensors connected and controlled via GPIO pins. It is developed and maintained by the Raspberry Pi Foundation and is released under the BSD 3-Clause "New" or "Revised" license. This license is very permissive: it allows redistribution of the library, both in source and binary format with or without modifications.

However, you must respect 3 constraints:

1. Redistribution in source must carry the copyright message throughout.
2. The redistribution in binary must carry the copyright message in all its parts: in the documentation and in any other material related to the project.
3. The name of the copyright holder (Raspberry Pi Foundation) and that of any contributor may not be used in any manner or form to endorse or promote products using the library without specific written permission.

Device Modeling

The key feature of the GPIOZero library is to provide a high-level model or interface for devices commonly used with the Raspberry Pi. Thus, unlike other libraries, it does not require the programmer to manage individual pins but provides an interface and a set of high-level methods specific to each supported component.

The components supported by version 1.4.1 are as follows:

- Led;
- Led with PWM brightness control;
- Board led;
- Led Bar;
- Led Traffic Light;
- RGB led;
- Button;
- HC-SR501 PIR Motion Sensor;
- LDR brightness sensor;
- HC-SR04 ultrasonic distance sensor;
- DC motor (direct current);
- Two-wheel robot (RC car);
- MCP300x Analog-to-Digital Converter (AC/DC);
- BlueDot;
- GPIO Remote Control;
- TCRT5000 Line Sensor;
- Acoustic Buzzer;
- SG90 servo motor.

In addition to these components, the library can support many other external devices through the extension of specific classes. In other words, the programmer maker can extend the library to support external devices not yet present in the list we have just given. We will now describe the use of the library to control two simple devices such as an LED (Output) and a button (Input). We will also compare it with the Arduino platform highlighting the advantages and the high level of expressiveness of the GPIOZero library.

LED

The led belongs to the class of output devices and is modeled by a simple intuitive interface. For example, the *on()* and *off()* methods allow the led to be turned on and off.

led.on()
led.off()

As you can see, these methods do not refer to a specific GPIO pin but are an abstraction of the actual characteristics of the LED. For example, a similar code for the Arduino platform could be:

digitalWrite(13, HIGH); // turn on the led
digitalWrite(13, LOW); // turn off the led

In this case the abstraction is of a lower level. The code sets the logical state of a specific GPIO and makes no reference to the use of the LED. Clearly if an LED is connected to GPIO 13 it will turn on and off. Both solutions are easy to understand but the abstraction provided by the LED class of the GPIOzero library is more intuitive and understandable for those who have just approached the world of Raspberry Pi programming. Another example of the intuitiveness provided by GPIOZero is the blink of an LED.

Another example of the intuitiveness provided by GPIOZero is the blink of an LED.

led.blink() #makes the led blink forever with a period of 1 second

The code for the Arduino platform would be:

for(;;) { // infinite loop
digitalWrite(13, HIGH); // turn on the led
delay(1000); // waits for 1000ms
digitalWrite(13, LOW); // turn off the led
delay(1000); // waits for 1000ms
}

Thus, through the use of the GPIOZero library, the maker-programmer can focus more on the logic of his control application, leaving the implementation details to the library itself.

BUTTON

The button belongs to the input device class and is modeled by a simple intuitive interface. For example, the *wait_for_press()* method puts the control program waiting for a button to be pressed.

button.wait_for_press() # wait until the button is pressed
print("Button pressed!") # execute the print instruction on the screen

The same functionality with Arduino can be implemented in this way:

for(;;) { // infinite loop
int button = digitalRead(7); // reads the logical state of GPIO 7 to which the button is connected
if (button == HIGH) // if the logical state of GPIO 7 becomes HIGH
Serial.println("Button pressed!"); // prints the string on the screen
}

Similarly to the LED also for the button the abstraction level is lower. This code reads the logical state of a GPIO and if it becomes HIGH it prints the default message. Comparing the two portions of code, the simplicity of use of the Button interface is immediately evident. In addition, GPIOZero allows the writing of a more compact and expressive code.

The *wait_for_release()* method places the program waiting for the release of a button.

button.wait_for_release() # wait until the button is released
print("Button released!") # execute the print instruction on screen

Here is the version for Arduino:

```
button_state = LOW; // initially the button is not pressed
for(;;) { // infinite loop
int button = digitalRead(7); // reads the logical state of GPIO 7
to which the button is connected
if (button == HIGH && button_state == LOW) //button has
been pressed
button_state = HIGH; //remember that the button is pressed
if (button == LOW && button_state == HIGH) //the button has
been released
Serial.println("Button pressed!");
}
```

As it is easy to see, for a slightly more complex functionality the Python code has remained similar if not the same as the previous case, while the code for the Arduino platform has become much more complicated. It is no longer of immediate understanding to the less experienced. Similar considerations can be made for all other devices supported by the GPIOZero library.

INSTALLING GPIOZERO

The GPIOZero library is distributed and installed, by default, along with the Raspbian Desktop operating system. Other operating systems, including Raspbian Lite, do not provide the library by default. However, you can install the GPIOZero library in any environment and operating system that supports Python. Assuming **python3** is already installed, installing GPIOZero is simple and involves a few steps as follows. If it is not present, we need to install the pip packet manager.

sudo curl https://bootstrap.pypa.io/get-pip.py | sudo python3

Now we can verify the correct installation by typing the command in the first line of the screenshot below. We should get something similar in response to what we see right after the command.

pip --version
pip 18.0 from /usr/local/lib/python3.5/dist-packages/pip (python 3.5)

To install the gpiozero library for python3, we can use the command:

sudo pip3 install gpiozero

And, of course, we can make sure that the installation of the GPIOZero library was successful by typing what we see in the following line.

pip3 list | grep gpiozero

If the output of the previous command is equivalent to what we see immediately below this line, we can consider gpiozero ready to be used.

gpiozero 1.4.1

In this chapter, we have introduced the python library GPIOZero, we have seen what are the steps and instructions to install it on Raspberry Pi, so we have acquired all the useful elements in order to start managing different types of external devices through the Raspberry. In the next chapters, we will describe how to use GPIOZero to control external devices and sensors.

Chapter 13

LED Control With GPIOZero - Part 1

In the introduction chapter, the GPIOZero python library was defined as follows: *a Python library, for Raspberry Pi, that provides a simple interface, for devices and sensors controlled via the GPIO.* As trivial as it may seem, **controlling an LED** is the equivalent of the "Hello World" program typical of any programming language. Therefore in this chapter is described, through practical examples, how to use the library GPIOZero to control a led. The methods provided by the library are also presented and analyzed.

LED

The LED is a polarized component. Its 2 pins are marked with a polarity sign (+ or -) so they are not equivalent.

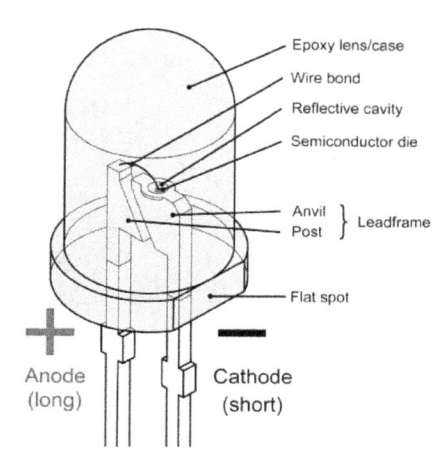

Looking at the led you will notice that the two pins have **different lengths**. This is very useful to be able to identify them correctly:

1. The longest pin, called the **anode**, corresponds to the + (plus) sign.
2. The shorter pin, called the **cathode**, corresponds to the - (minus) sign.

So, to make the LED light up, you need to connect the anode to the + terminal of a battery or power supply and the cathode to the - terminal of the battery or to the ground of the circuit.

LED class

The LED (Light Emitting Diode), within the GPIOZero library, belongs to the class of output devices (DigitalOutputDevice) and is modeled by a simple intuitive interface.

The constructor of the class has the following syntax:

LED(pin,
active_high=True,
initial_value=False,
pin_factory=None)

1. pin.
   It is a mandatory parameter of integer type that represents the pin number in BCM numbering to which the led is connected.
2. active_high.
   It is an optional boolean parameter (with default value True) that determines how the led is connected to the GPIO pin.
3. initial_value.
   It is an optional boolean parameter (with default value False) that determines the initial state of illumination of the led (so by default the led is off).
4. pin_factory.
   It is an optional parameter for advanced use only. It allows specifying a GPIO pin factory different from the one provided by GPIOZero.

The main methods of the class are as follows:

on()	Turn the LED on
off()	Turn the LED off
toggle()	It reverses the lighting state of the led. If it is on it turns it off, on the contrary, if it is off it turns it on.
blink(on_time=1, off_time=1, n=None, background=True)	It makes the LED blink repeatedly.

More information needs to be provided on the blink method.

- *on_time*: it is an optional parameter, float type (with default value 1), which represents the number of seconds the led is on.
- *off_time*: it is an optional parameter, float type (with default value 1), which represents the number of seconds the led is off.
- *n*: it is an optional parameter, integer type (with default value None), which represents the number of blinks of the led. By default the led blinks infinitely.
- *background*: it is an optional parameter, of type boolean (with default value True), that indicates if the method returns immediately to the main flow or if the method is blocking. By default, the method returns to the main stream immediately.

The main attribute of the class is *is_lit*, of type boolean, which represents the lighting status of the led. It takes value *True* if the led is lit or *False* if the led is off.

LIST OF MATERIALS

The following components and material are needed to make this circuit:

- 1x Raspberry Pi;
- 1x Breadboard;
- 1x 3 or 5 mm red led;
- 1x 220 Omh Resistor;
- 2x Rainbow wire.

Wiring

Before moving on to the python code it is necessary to assemble the electric circuit that connects the led to the Raspberry Pi. The figure displays the connections needed to control the led with the Raspberry Pi.

Follow the figure to make the following connections:

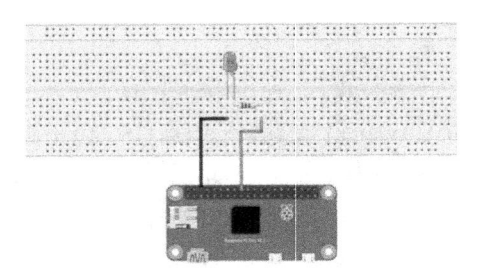

1. Connect with a wire (the color is not important) the cathode of the led with the ground pin of the Raspberry Pi that coincides with the physical pin number 6.
2. Connect the led anode with a 220 Omh resistor pin or rheophore.
3. Connect the free rheophore of the resistor with GPIO 24 of the Raspberry Pi which coincides with physical pin number 18.

Limiting Resistor

As shown in the connection diagram, the led is connected to the Raspberry Pi via a 220 Omh resistor. This resistor is called *limiting resistor* and has the task of limiting the electric current flowing through the pins of the Raspberry Pi and through the led. The limiting resistor is used to prevent the led from "burning out" and most importantly, to avoid possible damage to the Raspberry Pi circuits connected to the GPIO pin. In fact, the Raspberry Pi can supply a maximum of 16mA per GPIO up to a total maximum of 78mA.

The limiting resistor is sized according to the following formula:

Limiting resistor value = (Vin – Vfled) / Iled

1. **Vin**: it is the supply voltage of the led. In our case 3.3V;
2. **Vfled**: it is the specific forward voltage of the led. In the case of a red led it is about 1.8V;
3. **Iled**: the current flowing through the led.

So, in our example:

Limiting resistor value = (3.3 – 1.8) / 6mA = 250 Omh

A commercial resistor of 250 Omh does not exist, but you can safely use the 220 Omh resistor, making the LED and the Raspberry Pi flow about 6.8mA (well below the maximum value of 16mA).

Hello World

Now the led is properly connected to the Raspberry Pi. All that remains is to write the control code.

Step 1: Open the **Thonny** editor available on Raspbian. Obviously, any other editor will do. Thonny, however, is already present on Raspbian and is integrated with the python environment. So it allows python program execution directly from the editor.

Step 2: let's code!

```
from GPIOZero import LED #1
from time import sleep #2
print("Init Led") #4
led = LED(24) #5
print("Start blinking Led") #7
while True: #8
print("Turn on Led")
led.on() #10
sleep(1) #11
print("Turn off Led")
led.off() #13
sleep(1) #14
```

Line 1 – It imports the LED class which contains the class to control the LED connected to the Raspberry Pi;

Line 2 - It imports the sleep method that allows you to sleep for a specified amount of time;

Line 4 - It prints the string "Led initialization" to the Thonny console/shell;

Line 5 - It creates an object of type LED by connecting it to GPIO 24 and assigning it to led;

Line 7 - It prints the string "Start flashing Led" in the Thonny console/shell;

Line 8 - It starts an infinite cycle;

Line 10 - It turns on the led using the on() method;

Line 11 - It puts the program on hold for 1 second using the sleep method;

Line 13 - It turns off the LED using the off() method;

Line 14 - It puts the program on hold for 1 second;

The led is repeatedly switched on and off, for 1 second, generating a blink.

Step 3: Press the green run button in Thonny's icon bar.

Thonny will prompt you to specify the file name and save it before continuing execution. Therefore you must enter a valid file name in the text box and then press the *Save* button. At this point, if everything has been done correctly, Thonny will execute the LED flashing program.

The result you should get is the one in the figure (with flashing led).

In this chapter we have introduced the physical led describing its characteristics as well as described the class LED and its constructor. We have realized the simple circuit to connect the led to the Raspberry Pi, finally, we have also seen how to write the control code, with the GPIOZero library, to make the led flash. In the next chapter, we will present more code examples to control a led and we will analyze other useful methods of GPIOZero library.

Chapter 14

LED Control With GPIOZero - Part 2

We have introduced the physical **led** and described the class LED and its constructor in the previous chapter. We created a Hello World program to blink the led. Now we will see other code examples to control a led and we will analyze other methods of the **GPIOZero library**.

GPIO Current Sink

In the first part, we wired the led so that the current was supplied from the GPIO 24 pin and passed through the led to ground (pin 6). In this scenario, we say that GPIO 24 is *current source* i.e. it provides current. The GPIO of the Raspberry Pi can also operate in **current sink** mode that is absorbing current.

The figure displays the connections required to drive the LED with the GPIO in current sink mode.

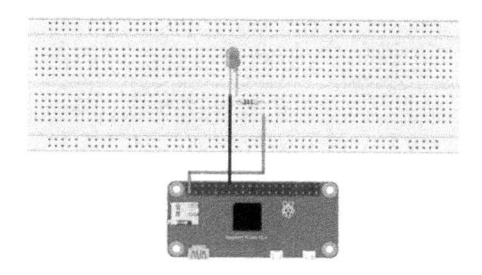

Follow the figure to make the following connections:

1. Connect with a wire (the color is not important) the cathode of the led with the GPIO pin 24 that coincides with the physical pin number 18.
2. Connect the led anode with a 220 Omh resistor pin or rheophore.
3. Connect the free resistor rheophore with VCC that is present on pin number 1 of the Raspberry Pi.

So, in this case, the current that feeds the led is supplied by pin 1 and absorbed by GPIO 24 (current sink) while in the previous wiring diagram the current is supplied by GPIO 24 and absorbed by ground pin number 6 (current source). In this mode, the control code is the following.

```
from GPIOZero import LED
from time import sleep
print("Init Led")
led = LED(24, active_high=False)
print("Start blinking Led ")
while True:
print("Turn on Led")
led.on()
sleep(1)
print("Turn off Led")
led.off()
sleep(1)
```

As you can see, the code is identical to the previous case except for the way the led object was created. In this case, the value *False* has been used for the *active_high* parameter of the constructor. This parameter changes the inner workings of the *on()* and *off()* methods but leaves the programming interface unchanged.

Blink With The toggle() Method

The Hello World or blinking led example is designed using the *on()* and *off()* methods to demonstrate its operation. However, the GPIOZero library offers other methods to achieve the same goal. The *toggle()* method, when invoked, reverses the state of the led, turning it on if off and off if on. So the Hello World code can be rewritten more compactly like this:

```
from GPIOZero import LED
from time import sleep
print("Init Led")
led = LED(24)
print("Start blinking Led")
while True:
    led.toggle()
    sleep(1)
```

Blink() method

The LED class provides the *blink()* method that performs the task of blinking the LED. However, to use the *blink()* method, we need to make some modifications to the previous code:

1. Remove the infinite while loop.
2. Replace *toggle()* with *blink()*.
3. Replace *sleep()* with *pause()*.

Here's the updated code:

```
from GPIOZero import LED
from signal import pause
print("Init Led")
led = LED(24)
print("Start blinking Led")
led.blink()
pause()
```

The *pause()* method allows the execution of the main stream to be placed on pause for an indefinite period. This is necessary because the *blink()* method, when invoked, creates a separate thread of execution and immediately returns control to the main stream. Thus, if the *pause()* method were not present, the program would terminate immediately and the led would turn off. There are several ways to make a led blink but, using the *blink()* method, the code is particularly compact.

The last example highlights, in terms of expressiveness and code density, the potential of the GPIOZero library. We have explored further methods that the LED class provides to control a led, described two fundamental concepts, namely current sink, and current source, and described the use of the *toggle()* method, then analyzed in detail the *blink()* method and all its call parameters. In the third and last part, we will complete the analysis of the LED class.

Chapter 15

LED Control With GPIOZero - Part 3

In the first and second part, we looked at important concepts such as current sink and current source. We have also described some methods of the LED class. In this third and last part, we will see more code examples to control an LED and we will analyze more methods of the GPIOZero library.

The *blink()* method by default generates a symmetrical blink in which the LED stays on and off for the same amount of time. To obtain an asymmetric blink, where the on time is different from the off time, it is necessary to use the *on_time* and *off_time* parameters of the blink method. The *on_time* parameter defines the time the led remains on while the *off_time* parameter defines the time the led remains off. The following code then blinks a led keeping it on for a tenth of a second every half second.

from GPIOZero import LED
print("Init Led")
led = LED(24)
print("Start blinking Led")
led.blink(on_time=0.1, off_time=0.4, background=False)
print("End blinking led")

The *pause()* method was not used because the *blink()* method was invoked with the *background = False* parameter. This means that the execution of the *blink()* method does not happen in a separate thread so the control will never be returned to the main stream. In this case, the call to the *blink()* method is called "blocking." As you can see in the console the string "*End blinking led*" is never printed because the last instruction executed is *blink()*.

Limited Blinking

Until now, in all the examples presented, the blinking of the led has always been infinite. There is a possibility to make the led blink for a finite number of times. The GPIOZero library provides a simple and effective way to achieve this.

from GPIOZero import LED
print("Init Led")
led = LED(24)
print("Start blinking Led")
led.blink(on_time=0.1, off_time=0.5, n=10, background=False)
print("End blinking led")

The *blink()* method was invoked by specifying the parameter n=10. This means that the led will blink exactly 10 times and then be turned off. Executing this code, at the end of the blink, the string "End blinking led" is printed, because of the *blink()* method, although blocking, returns the control to the main stream after 10 blinks.

In this third part, we have completed the analysis of all the methods that the LED class provides to control an LED and we have also understood how to make it blink in an asymmetric and limited way. In the next chapter, we will describe the use of the GPIOZero library to control an input device such as, for example, a simple button.

Chapter 16

Digital Input: Pull-Up and Pull-Down

In this chapter, we address the concepts of **Pull-Up and Pull-Down resistors** needed to implement a **Digital Input** with the Rapsberry Pi. Often when you want to read the state of a digital input, for example, a button connected to a GPIO on Rapsberry, unexpected effects occur: the button is read as "pressed" even though it actually isn't, or the state of the digital input quickly switches between *HIGH* and *LOW*. All these problems have a specific reason. Let's analyze the reasons why this happens and, above all, find out how to avoid it.

High Impedance

As we have already mentioned in some previous chapters the GPIO of the Raspberry Pi can be set to OUTPUT or INPUT mode. In the latter case, the pin in question is brought into a **high impedance state**. In short, it behaves as if it were electrically disconnected from the rest of the circuit. At its ends, there is not a precise voltage value but the voltage depends on the circuit to which the pin is connected. We can imagine as if in series to the pin is present a *100 MOhm* resistor. In this way, the *INPUT* pin requires a very small current to change its logic state. In other words, when a GPIO is in *INPUT* mode, if not correctly connected, it behaves as a "simple antenna" that can pick up any kind of noise or electrical/electromagnetic perturbation, changing its logic state in a random and uncontrollable way. This feature allows us to understand that to correctly read the state of a button, without spurious readings, it is necessary to connect it to the Raspberry Pi correctly.

Pull-Up and Pull-Down Resistors

This can be solved by appropriately connecting a resistor to the pin and button in question.

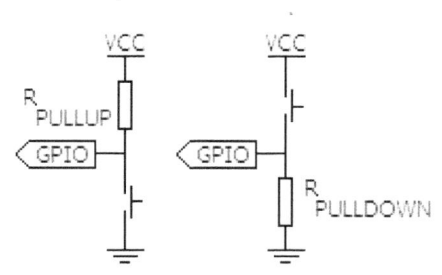

There are two possible configurations named according to the position of the resistor:

1. **Pull-Up**: when a resistor (10-100K Ohm) is inserted between the GPIO and the power supply (VCC).
2. **Pull-Down**: when a resistor (10-100K Ohm) is inserted between the GPIO and ground (GND).

Let's take a closer look at how it works.

PULL-UP

When the button is not pressed (*OFF*) the pin assumes stable logic state 1 (*HIGH*). Contrary to the case in which the resistor is not present the logic state 1 remains at the GPIO input of the Raspberry Pi in a stable and permanent way. Once the button is pressed the current flows to the ground bringing the pin to logic state 0 (*LOW*).

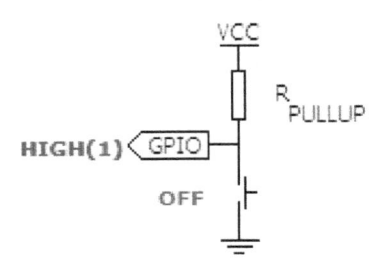

PULL-DOWN

With the *OFF* button (not pressed) the GPIO assumes stable logic state 0 (*LOW*). Since the resistor is present the logic state 0 remains at the input of the Raspberry Pi pin without oscillations. In this case when the button is pressed the current flows to the ground bringing the pin to logic state 1 (*HIGH*).

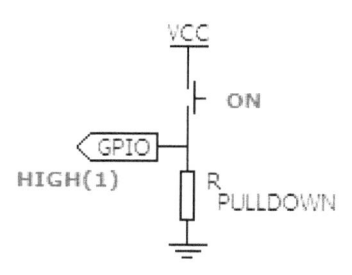

In conclusion, we can say that when connecting an INPUT device to the Raspberry PI **it is always necessary to use a Pull-Up or Pull-Down resistor to avoid wrong readings**. Fortunately, each GPIO of the Raspberry Pi is internally equipped with a Pull-Up or Pull-Down resistor that can be activated via software. In this chapter, we have addressed the age-old problem of spurious readings of a digital input. In the next chapter, we will apply what we learned by managing the reading of a button connected to the Raspberry Pi using the GPIOZero library.

Chapter 17

Button with GPIOZero - Part 1

We introduced the fundamental concept of pull up resistor and pull down resistor for the proper use of an input device with the Raspberry Pi in the previous chapter. In this chapter, we will see **how to use the GPIOZero library to read the state of a button** (input device) and we will do a detailed analysis of the Button class helping with practical examples in Python.

Button Class

The *Button* within the GPIOZero library belongs to the Input device class (*DigitalInputDevice*) and is modeled by a simple intuitive interface.

The constructor of the class has the following syntax:

*Button(pin, \
pull_up=True, \
bounce_time=None, \
hold_time=1, \
hold_repeat=False, \
pin_factory=None)*

Let's go into more detail about what we used.

- **pin**.
  Mandatory parameter of integer type, it represents the pin number in BCM numbering to which the button is connected.
- **pull_up**.
  Optional boolean parameter, it enables the internal *Pull-Up* (*True - Default*) or *Pull-Down* (*False*) resistor.
- **bounce_time**.
  Optional parameter of float type, it avoids software bounces by specifying the number of seconds in which changes in the state of the button are ignored, by default it is disabled.
- **hold_time**.
  Optional parameter of float type, it specifies the duration in seconds of the button pressure after which the *when_held* handler is executed.
- **hold_repeat**.
  Optional parameter of boolean type, if *True* (*Default*) every *hold_time* seconds it calls the *when_held* handler, if *False* it calls it only once.
- **pin_factory**.
  Optional parameter for advanced use only, it allows specifying a GPIO pin factory different from the one provided in GPIOZero.

The most important methods of the class are the following:

wait_for_press(Timeout=None)	Blocking call that waits for the button to be pressed or for the float type Timeout to expire.
wait_for_release(Timeout=None)	Blocking call that waits for the release of the button or the expiration of the float type Timeout.

The main events in this class are:

when_held	It performs the associated function when the button is held down for *hold_time* seconds.
when_pressed	It performs the associated function when the button is pressed.
when_released	It performs the associated function when the button is released.

Now we'll put what we've learned into practice with some simple examples of using a button.

PULL-UP CONFIGURATION

Let's make the button connections as in the figure.

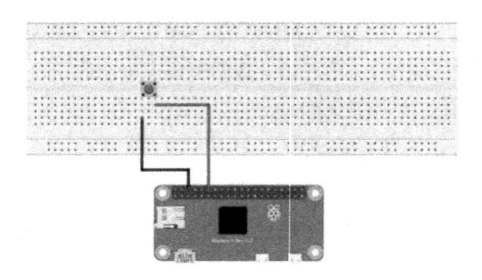

from gpiozero import Button #1
from time import sleep #2
print("Init button") #4
button = Button(18, pull_up=True) #5
while True: #7
if button.is_pressed: #8
print("Button pressed!") #9
else: #10
print("Button not pressed!") #11
sleep(0.5) #12

This code at line 5 initializes the button connected to GPIO 18 and sets the internal Pull-Up resistor. Then, as you can see in the figure, the other pin of the button should be connected to the GND pin of the Raspberry Pi. In line 8, by reading the is_pressed class property, we can understand if the button is pressed or not. Based on the returned result, a simple message is printed on the screen.

PULL-DOWN CONFIGURATION

We now realize the configuration with the Pull-Down resistor by connecting the button as shown in the figure.

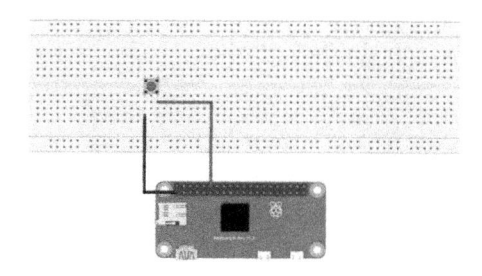

```
from gpiozero import Button #1
from time import sleep #2
print("Init button") #4
button = Button(18, pull_up=False) #5
while True: #7
if button.is_pressed: #8
print("Button pressed!") #9
else: #10
print("Button not pressed!") #11
sleep(0.5) #12
```

The code is the same as the previous one except for the constructor on line 5. In this case, pull_up is set to False, in fact, this sets on GPIO 18 the Pull-Down resistor. The operation is the same as the previous case.

In this chapter and in the previous one we have learned the fundamental concepts and faced the use of Pull-Up and Pull-Down resistors. We have described the Button class of the GPIOZero library and we have analyzed some control codes that take advantage of the methods and events provided by the class itself. In the next chapter, we will complete the analysis of the Button class and its methods.

Chapter 18

Button with GPIOZero - Part 2

We have seen how to connect and manage a button with the Raspberry Pi in the previous chapter as well as analyzed the Pull-Up, Pull-Down modes and methods of the Button class. In this chapter, we complete the detailed analysis of the methods of the Button class with the help of practical examples in Python.

Led Control with Button

In this case, we modify the circuit as shown in the figure by connecting a led to GPIO 23 of the Raspberry Pi.

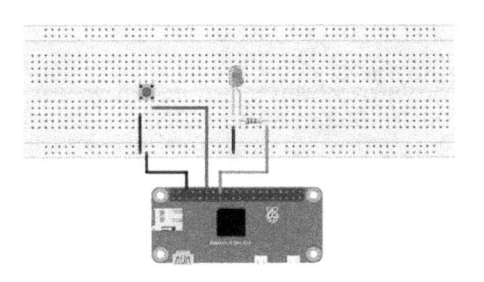

```
from gpiozero import LED, Button #1
from time import sleep #2
print("Init Led") #4
led = LED(23) #5
print("Init button") #7
button = Button(18, pull_up=True) #8
while True: #10
if button.is_pressed: #11
print("Button pressed - Led on!") #12
led.on() #13
else: #14
print("Button not pressed - Led Off!") #15
led.off() #16
sleep(0.5)
```

If at line 11 the button is pressed, then we turn on the led at line 13 otherwise we turn it off at line 16. The GPIOZero library allows us to obtain the same result in a more stringent way by using the *when_pressed* and *when_released* events.

```
from gpiozero import LED, Button #1
from signal import pause #2
print("Init Led") #4
led = LED(23) #5
print("Init button") #7
button = Button(18, pull_up=True) #8
button.when_pressed = led.on #10
button.when_released = led.off #11
pause()
```

In this case, we have linked the event of pressing the button (*when_pressed*) with the execution of the *on()* function of the led, while the event generated by the release of the button is linked to the execution of the *off()* method of the led. We get the same result as in the previous code but in a more synthetic way. There is a third way, even more compact, that allows to logically link the logical state of the button as input of the led.

```
from gpiozero import LED, Button #1
from signal import pause #2
print("Init Led") #4
led = LED(23) #5
print("Init button") #7
button = Button(18, pull_up=True) #8
led.source = button.values #10
pause()
```

In this case, the control value of the led via *led.source* is driven by the logical state of the button *button.values*.

Long Press Button

By handling the *when_held* event we now create a code that turns the LED on or off only if the button is pressed for more than 2 seconds.

```
from gpiozero import LED, Button #1
from signal import pause #2
print("Init Led") #4
led = LED(23) #5
print("Init Pulsante") #7
button = Button(18, pull_up=True, hold_time=2) #8
button.when_held = led.toggle #10
pause()
```

As you can see, on line 10, we used the toggle() method of the LED class to reverse the LED's on state.

In this chapter and the previous one, we have learned some basic concepts and observed the use of Pull-Up and Pull-Down resistors. We have also described the Button class of the GPIOZero library and analyzed various control codes that take advantage of the methods and events provided by the class itself. In the next chapter, we're going to drive by button a load that cannot be directly connected to the Raspberry Pi.

Chapter 19

GPIOZero Relay and Library

In a previous chapter, we learned how to handle a load directly connected to a GPIO on the Raspberry Pi. However, for purely technical and electrical reasons, it is not always possible to do this for any load. **We will therefore learn how to drive an external load via a Relay.**

Relay

A Relay is an electrically controlled switch that is used when it is necessary to drive a high power load via a low power signal, such as a GPIO on the Raspberry Pi.

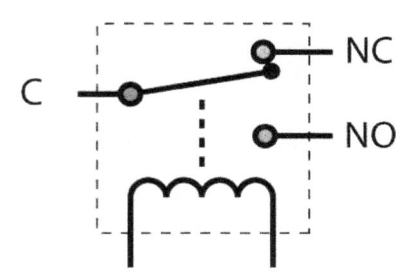

The high power load is connected between the *COMMON* and *NORMALLY OPEN* poles. For example, let's imagine that we want to drive, as we said, a common 220V lamp. This certainly cannot be connected directly to the Raspberry Pi. The Relay can be used to drive the lamp via a GPIO pin. A few milliAmps can then drive a 220V lamp of 100W.

RELAY MODULE

In this chapter, we introduce the use of the 5V Relay module (very common in the maker community using Raspberry Pi). The module, which contains a Relay that can be controlled at 5V, can be used for loads with maximum current rating of 10A, both DC and AC. However, it is highly recommended never to reach this value for safety reasons. If you use loads with maximum currents of 50/60% of the nominal value you are sure to work safely.

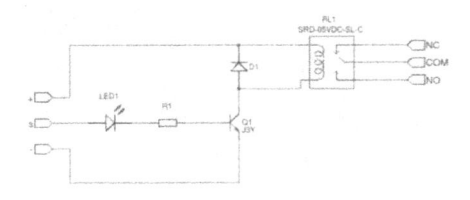

The module, in addition to the Relay, also includes a few other passive components, useful to make it usable and controllable via a GPIO. This module has 3 pins:

- **+**: pin for 5V power connection;
- **-**: Ground or GND connection pin;
- **S**: Relay control pin.

The control pin S is connected, via a resistor, to the base of an NPN transistor. This solution allows to excite the coil of the Relay with a few mA of current provided by a GPIO of the Rapsberry Pi.

Direct Control with Raspberry Pi

Let's now make the circuit to properly connect the Relay module to the GPIO #24 of the Raspberry.

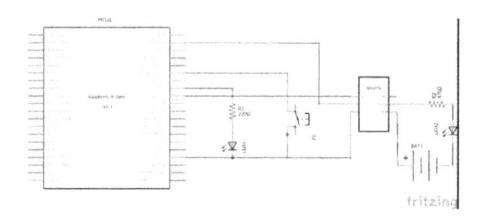

We need the following materials:

- Raspberry Pi;
- 5V Relay Module;
- One button;
- Two LEDs;
- One 220 Ohm resistor;
- One 470 Ohm resistor;
- One 9V transistor type battery.

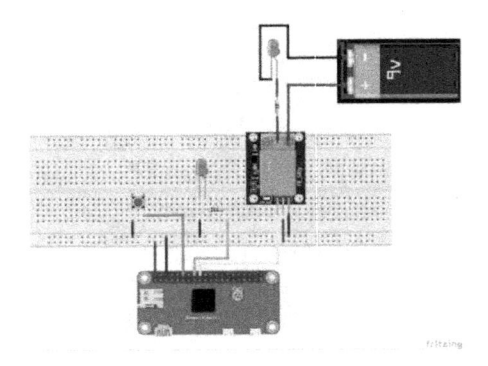

Normally a power load is connected to the ends of the Relay but for practical reasons, in this circuit, a normal led is used. However, the functioning remains unchanged even when the connected load changes. Once the button is pressed for at least half a second, the Relay is activated and the load is powered. At the same time, the first led lights up to indicate that the high power load is active. Vice versa, pressing again the button for at least half a second, the Relay switches to interrupt the supply of the 220V lamp (simulated by the LED). The control led will turn off. Once the electrical connections are made, run this Python code:

from gpiozero import LED, Button, OutputDevice
from signal import pause
def drive_load (): #4
led_control.toggle()
rele.source = led_control.values #6
print("Init Led")
led_control = LED(23) #9
relay = OutputDevice(24) #10
print("Init button")
button = Button(18, pull_up=True, hold_time=0.5) #13
button.when_held = drive_load #15
pause()

The code is very simple, it implements and reuses concepts already described in previous chapters. The main novelty is the use of the OutputDevice class to model the Relay. In fact, the GPIOZero library does not have a specific class for the Relay module, as it does for the LED and the Button.

- At line 9 we initialize the control led connected to GPIO #23;
- At line 10 we define the relay connected to GPIO #24;
- At line 13 we define the button connected to GPIO #18 by setting a pull-up resistor and a pressure time of 0.5 seconds;
- Finally at line 15 we connect to the event of the button *when_held* the execution of the function *drive_load* previously defined at line 4.

Note how the state of the Relay is set at line 6 according to the value of the control LED which in turn is determined by the state of the button. In this way, we have implemented the desired operation for controlling the Relay and the load connected to it.

Indirect Control with Raspberry Pi

In the previous example, we saw how to directly connect the Relay module to the Raspberry Pi. There is another method that is useful in case you want to directly drive a single Relay (not Relay module). In this alternative example, we will use an NPN transistor interposed between the GPIO of the Raspberry and the Relay module.

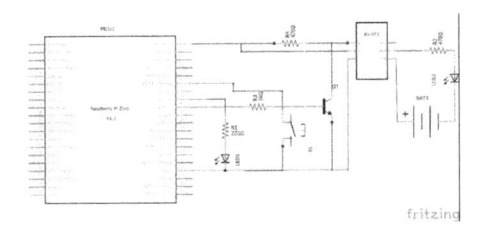

For the circuit in question we need the following material:

- Raspberry Pi;
- 5V Relay Module;
- One button;
- Two LEDs;
- One 220 Ohm resistor;
- One 470 Ohm resistor;
- One 1K Ohm resistor;
- One common NPN transistor (any abbreviation will do);
- One 9V battery of transistor type.

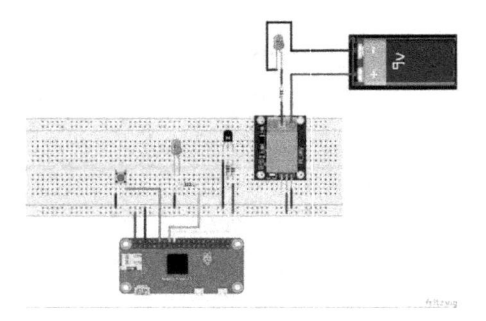

As you can see from the connection diagram, GPIO #24 in this case is connected via the 1K Ohm resistor to the base of the NPN transistor. Pin S of the Relay module is connected to the collector of the transistor and, via a 470 Omh resistor, to VCC (5V). So, in this way, the current entering the Relay module is no longer supplied directly from GPIO #24 but from the supply line through the NPN transistor. Let's see now the code related to this case:

```
from gpiozero import LED, Button, OutputDevice
from signal import pause
def drive_load (): #4
led_control.toggle()
rele.source = led_control.values #6
print("Init Led")
led_control = LED(23) #9
relay = OutputDevice(24, active_high=False) #10
print("Init button")
button = Button(18, pull_up=True, hold_time=0.5) #13
button.when_held = drive_load #15
pause()
```

The main difference is the initialization of the Relay at line 10. In this case, by setting *active_high=False*, the GPIO will work in reverse logic. That is, it will be *LOW* when it is *on()*, and *HIGH* when it is *off()*. It is necessary to move in this way because we have used an NPN type transistor that realizes an inverting function (NOT). So with the GPIO in reverse logic, in series with the inverting NPN, we get the expected operation and the circuit works exactly as in the first case, so the Relay is controlled by pressing the button.

In this chapter, we have seen how to use the Raspberry Pi to drive a high power load (a lamp, a fan, an electrical outlet) that technically cannot be connected directly to the GPIOs of the Raspberry. We have described, two different ways of connection, one direct and one indirect, through the use of a transistor. The techniques learned can be used in a myriad of projects, especially in home automation, where the Raspberry Pi is used to drive any home device.

Chapter 20

Led Brightness Control With GPIOZero and PWM

In chapter 13 we introduced the use of the LED and learned how to drive it. In particular, we described the use of the *on()* and *off()* methods to turn the led totally on and off. In this lesson, we will illustrate the use of a technique called **PWM** to control the brightness of a led.

Pulse Width Modulation

You don't have to go far to get a convincing definition, so here's how Wikipedia defines the topic at the time I'm writing this chapter: "*Pulse width modulation (PWM), or pulse-duration modulation (PDM), is a method of reducing the average power delivered by an electrical signal, by effectively chopping it up into discrete parts. The average value of voltage (and current) fed to the load is controlled by turning the switch between supply and load on and off at a fast rate. The longer the switch is on* compared to the off periods, the higher the total power supplied to the load."

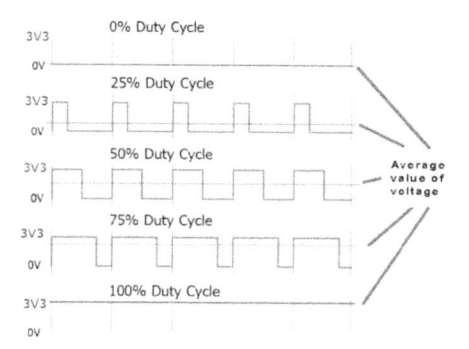

```
3V3          0% Duty Cycle
0V
             25% Duty Cycle
3V3
0V
             50% Duty Cycle
3V3                                          Average
0V                                           value of
                                             voltage
             75% Duty Cycle
3V3
0V
             100% Duty Cycle
3V3
0V
```

The figure shows the time course of the output of a PWM controlled GPIO pin. As you can see the output can be in the logic state HIGH (3.3 volts) or in the logic state LOW (0 volts). What varies is the **duty cycle**, that is the percentage ratio between the time the output is HIGH and the time the output is LOW. Since the voltage alternates between the two values 0 and 3.3V we can calculate what will be the average voltage on the GPIO pin. In the figure, this voltage is represented by a horizontal line. It directly depends on the *duty cycle* value. If *duty cycle* is 25% then average voltage at GPIO pin will be 0.825V (25% of 3.3V). If the *duty cycle* is 75% then the average voltage will be 2.475 volts (75% of 3.3V). The same reasoning applies for any *duty cycle* value in the range 0-100%. Therefore, since the brightness of the led is proportional to the current flowing through it, by varying the voltage at its ends, with a constant resistance, we will obtain the lighting of the led with an intensity value proportional to the *duty cycle*.

PWMLED Class

A variable brightness led (PWM LED), within the GPIOZero library, belongs to the Output device class (DigitalOutputDevice) and is modeled by a simple intuitive interface. The constructor of the class has the following syntax:

PWMLED(pin,
active_high=True,
initial_value=0,
frequency=100,
pin_factory=None)

Let's detail some of the elements in the following list.

- *pin*: mandatory parameter, integer type, it represents the pin number in BCM numbering to which the led is connected.
- *active_high*: optional boolean parameter, default value *True*, determines the connection mode of the led to the GPIO pin.
- *frequency*: optional integer parameter, default value 100, it declares the frequency in Hertz of the wave generated on GPIO pin.
- *pin_factory*: optional parameter for advanced use only. It allows specifying a GPIO pin factory different from the one provided by GPIOZero.

Compared to the LED class, a new method is added that turns the LED on and off with a *fade-in* and *fade-out* effect:

pulse(fade_in_time=1, fade_out_time=1, n=None, background=True)

- *fade_in_time*: time in seconds for *fade-in*, i.e. to go from 0% to 100% brightness.

- *fade_out_time*: time in seconds for *fade-out*, i.e. to go from 100% brightness to 0% brightness.
- *n*: number of fades, *None* means unlimited number of fades.

Variable Brightness LED

Let's make a circuit similar to the one in chapter 14.

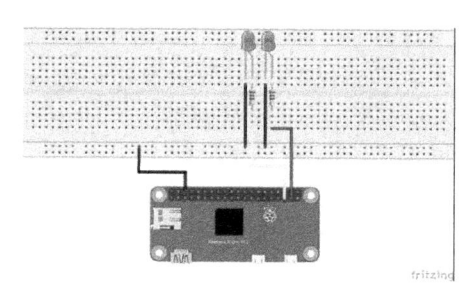

Here are some examples of using the class:

from gpiozero import PWMLED
from time import sleep
from signal import pause
pwm_led1 = PWMLED(19)
pwm_led2 = PWMLED(26)
pwm_led1.value = 1
pwm_led2.value = 0
pause()

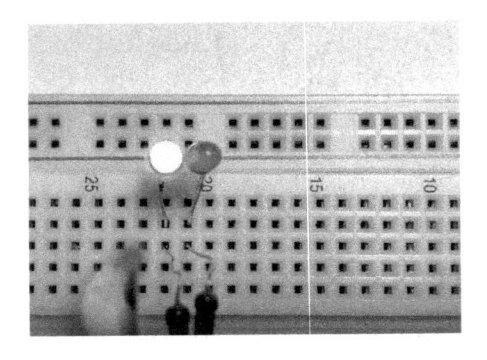

As you can see in the picture the first led is completely on while the second led is off. The float value passed to the value parameter represents the percentage of brightness of the led (0 = 0%, 1 = 100%).

from gpiozero import PWMLED
from time import sleep
from signal import pause
pwm_led1 = PWMLED(19)
pwm_led2 = PWMLED(26)
pwm_led1.value = 1
pwm_led2.value = 0.1
pause()

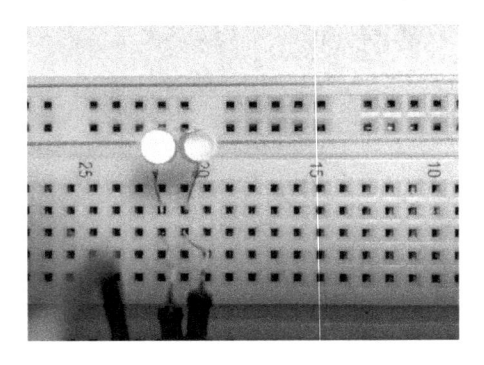

Now the first led is on with 100% brightness while the second led is on with 10% brightness (0.1).

```
from gpiozero import PWMLED
from time import sleep
from signal import pause
pwm_led1 = PWMLED(19)
pwm_led2 = PWMLED(26)
while True:
for val in range(0,101):
pwm_led1.value = val * 0.01
pwm_led2.value = 1 - (val * 0.01)
sleep(0.2)
sleep(1)
```

In the latter example, the first LED will gradually turn on with a fade-in effect while the second will turn off with a fade-out effect.

In this chapter, we introduced the concept of PWM modulation and learned how to use it with the Raspberry Pi to vary the brightness of a led. This concept is very important as it applies in many other situations (e.g. speed control of a DC motor).

Chapter 21

Led Bar with GPIOZero

In this chapter, we will illustrate how to control a **led bar** with Raspberry Pi and the GPIOZero library. A led bar is an array of leds that is generally used to display a level (ex: battery level, audio level). The led bar can be made as an already assembled component, or as we do in this chapter, it can be made from an array of **discrete leds**.

Led Bar Circuit

To make the circuit of the led bar, we need 5 leds, 5 resistors of 100 Ohm, and some connecting wires. Obviously, we can create shorter or longer led bars, removing or adding led.

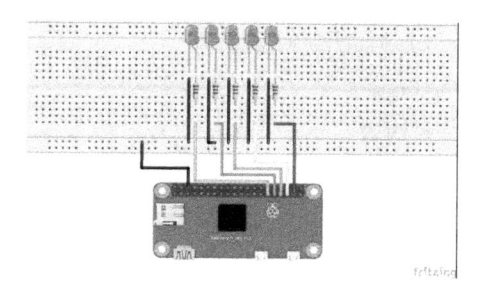

Led Class Bar Graph

The led bar, within the GPIOZero library, belongs to the composite output device class (*CompositeDevice*) and is modeled by a simple intuitive interface. The constructor of the class has the following syntax:

*LEDBarGraph(*pins, initial_value=0)*

- *pins*: set of GPIO pins to which the various leds of the led bar are connected;
- *initial_value*: initial value between *-1* and *+1*, default set to zero (*led bar off*).

The LEDBarGraph class provides *on()*, *off()* and *toggle()* methods that work the same way as Led class, but applied to the whole array. Let's see a code example in which we make the led bar blink with period 1 second.

from gpiozero import LEDBarGraph
from time import sleep
led_bar = LEDBarGraph(5, 6, 13, 19, 26)
while True:
led_bar.on()
sleep(1)
led_bar.off()
sleep(1)

As you can see, the **blink** of the **led bar** is achieved in the same way as the LED blink. Only the constructor changes as we use the class ***LEDBarGraph*** instead of the class **LED**. As mentioned above, the GPIOs, in BCM numbering, to which the LEDs of the led bar are connected, are passed to the constructor of the class.

Value Property

Also, this class exposes the value property. For the *Led* class it could assume value *0 (off)* or *1 (on)*. For the *LedBarGraph* class, this property can assume a *float* value between *0* and *1* or between *0* and *-1*. Values between 0 and 1 turn on in a proportional way (0% - 100%) the led bar from left to right. On the contrary, values between 0 and -1 from left to right. So, in our example with 5 leds, the first led represents 20% of the led bar, so to turn it on it is necessary to assign to value the value 0.2 or 1/5. Similarly to turn on the 60% led bar it is necessary to assign to value 0.6 or 3/5. Let's see now the control code of the led bar that turns on all the bar, led by led from left to right and then from right to left.

```
from gpiozero import LEDBarGraph
from time import sleep
NUM_LED=5
led_bar = LEDBarGraph(5, 6, 13, 19, 26)
while True:
    for led in range(1,NUM_LED+1):
        led_bar.value = led/NUM_LED
        sleep(0.2)
    for led in range(1,NUM_LED+1):
        led_bar.value = -led/NUM_LED
        sleep(0.2)
```

So, we can generalize by saying that to turn on the first *N* leds of a bar of *M leds*, we must assign to value the value *N/M*.

In this chapter, we learned how to use a very useful class to control a led bar. This type of led can be very useful in many maker projects to be done with the Raspberry Pi.

Chapter 22

RGB LED with GPIOZero

In this chapter, we describe how to drive an RGB LED with the GPIOZero library.

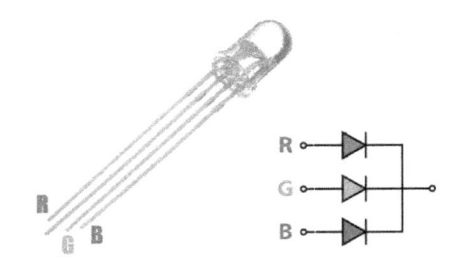

A RGB led (Red Green Blue) is composed of 3 distinct leds enclosed in a single container. As you can see in the figure we can consider it as the union of a red led, a green led and a blue led. All leds have a common pin so RGB led has 4 different pins.

Shared Anode and Cathode

Two distinct configurations are possible depending on which pin the 3 LEDs have in common.

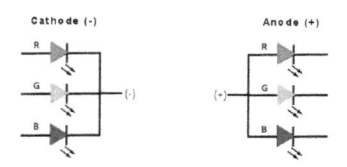

If the red, green, and blue LEDs have the minus (-) pin in common, this configuration is called common cathode. On the contrary, if the common pin is the plus (+), the configuration is called common anode. Therefore in the common anode configuration, the pins of the RGB led assume the meaning shown below.

- R: cathode of red led;
- G: cathode of green led;
- B: cathode of blue led;
- +: anode common to all three leds.

Vice versa, in the common cathode configuration, the pins of the RGB led are to be considered as immediately below.

- R: red led anode;
- G:green led anode;
- B: anode of blue led;
- -: cathode common to all three leds.

Limiting Resistor

The RGB led is composed by three distinct leds so three limiting resistors will be needed, one for each led. The leds have different threshold voltages, the resistors must be sized appropriately using the formula seen in the chapter *LED control with GPIOZero - part 1*. In our example, considering that the GPIO of the Rapsberry Pi is 3.3V, we can use a 100 Ohm resistor for the red led, 22 Ohm for the green led, and 10 Ohm for the blue led.

RGBLED Class

The RGB LED (*Red Green Blue Light Emitting Diode*), within the GPIOZero library, belongs to the output device class (*DigitalOutputDevice*) and is modeled by a simple intuitive interface. The constructor of the class has the syntax you can see below.

RGBLED(red, green, blue,
active_high=True,
initial_value=(0, 0, 0),
pwm=True,
pin_factory=None)

- *red*: mandatory parameter, type *int*. The GPIO number connected to the red component of the led;
- *green*: mandatory parameter, type *int*. The GPIO number connected to the green component of the led;
- *blue*: mandatory parameter, type *int*. The GPIO number connected to the blue component of the led;
- *active_high*: optional parameter, *bool* type, default *True* for common cathode rgb led, False for common anode led;
- *initial_value*: optional parameter, *tuple* type, default *black*. The initial color when the led is turned on;
- *pwm*: optional parameter, *bool* type, default *True*. Sets the led to variable brightness;

- *pin_factory*: for advanced use only. It allows to specify a different GPIO pin factory than the one provided by GPIOZero.

Wiring Diagram

Before moving on to the python code it is necessary to assemble the electric circuit that connects the RGB led to the Raspberry Pi. The figure displays the connections needed to drive the led with the Raspberry Pi.

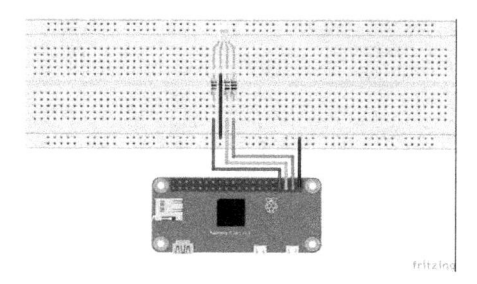

Let's see now a first example of code that makes the led flash using its three main colors.

```
from gpiozero import RGBLED
from time import sleep
RED = (1, 0 ,0)
GREEN = (0, 1, 0)
BLUE = (0, 0, 1)
rgb_led = RGBLED(13, 19, 26, pwm = False)
while True:
rgb_led.color = RED
sleep(1)
rgb_led.color = GREEN
sleep(1)
rgb_led.color = BLUE
sleep(1)
```

We have defined three tuples that represent the three colors *RED*, *GREEN*, and *BLUE*. The object *rgb_led* is created by instantiating the class *RGBLED* to which we passed the GPIO number in BCM numbering relative to the respective pins of the led. Note that we set the pwm parameter to *False* because in this example we don't want to get colors as a combination of the three main colors. Defining colors as a tuple of three numbers is quite awkward and laborious. The *colorzero* library helps us to simplify things. Let's install it with the following command:

pip install colorzero

This library allows us to specify the color of the led using some conventional names. For example, the red color is defined as *Color('red')* while the yellow color is defined as *Color('yellow')*. Now let's see a code similar to the previous one that lights the led with various colors using the *colorzero* library.

```
from gpiozero import RGBLED
from colorzero import Color
from time import sleep
rgb_led = RGBLED(13, 19, 26, pwm = True)
while True:
rgb_led.value = Color('red')
sleep(1)
rgb_led.color = Color('green')
sleep(1)
rgb_led.color = Color('blue')
sleep(1)
rgb_led.color = Color('yellow')
sleep(1)
rgb_led.color = Color('magenta')
sleep(1)
rgb_led.color = Color('aqua')
sleep(1)
```

As you can see in this case the constructor *pwm* parameter is set to *True*. It is necessary because we want to light the led with colors that are obtained from the combination of the 3 main colors. In the next example, we see how to blink an RGB led.

from gpiozero import RGBLED
from colorzero import Color
from signal import pause
rgb_led = RGBLED(13, 19, 26, pwm = True)
rgb_led.blink()
pause()

As you can see also in the case of the RGB led we use the *blink()* method which demonstrates once again the clarity and strong intuitiveness of the GPIOZero library. Let's see now how to blink between two different colors.

from gpiozero import RGBLED
from colorzero import Color
from signal import pause
rgb_led = RGBLED(13, 19, 26, pwm = True)
rgb_led.blink(on_color=Color('red'), off_color=Color('green'))
pause()

We have always used the *blink()* method but passing two additional parameters to specify the first color (*on_color*) and the second color (*off_color*) of the "blink." Even more interesting is the use of the *pulse()* method which allows us to create a "blink" between two colors in a continuous manner. In other words, the led will not be on red and then on green, as in the case of the blink, but it will vary gradually assuming all intermediate colors between red and green.

from gpiozero import RGBLED
from colorzero import Color
from signal import pause
rgb_led = RGBLED(13, 19, 26, pwm = True)
rgb_led.pulse(on_color=Color('red'), off_color=Color('green'))
pause()

In this chapter, we have seen Python examples, that even a simple RGB LED offers us many possibilities. We can easily obtain a series of luminous effects. Obviously, we will obtain even better visual results if we use several RGB LEDs working together.

Chapter 23

PIR Motion Sensor with GPIOZero

In this chapter, we describe how to detect motion using a **PIR sensor** and the GPIOZero library. All objects and living beings can emit energy in the form of light radiation in the infrared spectrum therefore invisible to the naked eye. Therefore a passive infrared sensor or PIR (*Passive InfraRed*) is an electronic device that measures the infrared rays (*IR*) radiated in its field of action. The most classic PIR sensor is the one that is normally used in home alarm systems.

PIR Sensor

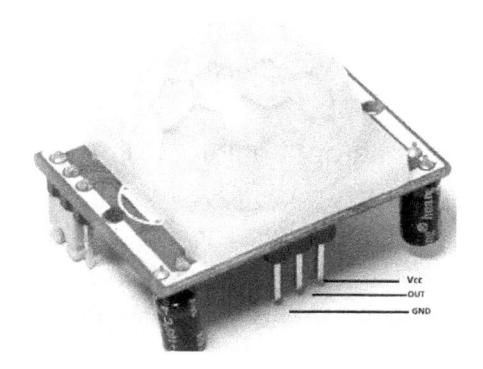

A typical PIR module for Raspberry Pi consists of the sensor itself and a control circuit with three pins:

- *VCC*: power supply pin, normally 5V;
- *OUT*: digital pin representing the sensor output ("present" detection or free field);
- *GND*: ground pin.

MotionSensor Class

The *PIR* sensor, within the GPIOZero library, belongs to the input device class (*SmoothedInputDevice*). The constructor of the class has the following syntax:

MotionSensor(pin=None,
pull_up=False,
active_state=None,
queue_len=1,
sample_rate=10,
threshold=0.5,
partial=False,
pin_factory=None)

- *pin*: is a mandatory integer parameter that represents the BCM pin number to which the PIR sensor is connected;
- *pull_up*: is an optional boolean parameter that enables the internal Pull-Down (*False* by default) or Pull-Up (*True*) resistance;
- *active_state*: is an optional boolean parameter. Useful in case you set pull_up to *None*. If set to *True* the state of the software pin follows the state of the hardware pin (e.g. hardware pin *True* implies software pin *True*). If set to *False* the state of the software pin is opposite to the state of the hardware pin (e.g. hardware pin *True* implies software pin *False*). If the *pull_up* parameter is set to a boolean value, then *active_state* is automatically set to the correct value;
- *queue_len*: is an optional integer parameter that represents the length of the queue used to store the passed sensor readings. By default, it is set to *1* (in practice the queue is disabled). In case the *PIR* sensor is unstable it may be useful to increase the value of this parameter;
- *sample_rate*: is an optional integer parameter that represents the number of readings per second operated by the sensor. By default, it is set to 100. Normally it is not necessary to change this parameter;
- *threshold*: is an optional float parameter that represents the threshold, calculated on the average of the values present in the internal queue, beyond which the sensor is considered active. Normally it is not necessary to modify this parameter;
- *partial*: is an optional boolean parameter set to *False* by default. When set to *False* the value of the *is_active* property is returned only if the internal queue is full (blocking reading). If it is set to *True* the reading is non-blocking, therefore the *is_active* value is calculated only with the

values present in the internal queue without waiting for it to fill up;
- *pin_factory*: is an optional parameter for advanced use only. It allows specifying a GPIO pin factory different from the one provided by GPIOZero.

The most important methods and properties of the class are as follows:

is_active	It returns the state of the PIR sensor according to the *active_state* parameter set in the manufacturer.
wait_for_motion(timeout=None)	It suspends the program until the sensor becomes *active* or the *timeout* has expired.
wait_for_no_motion(timeout=None)	It suspends the program until the sensor becomes *inactive* or the *timeout* has expired.
motion_detected	It returns a *True* value if the sensor has detected a movement otherwise it returns *False*.
when_motion	Function that is executed when the sensor changes state from *inactive* to *active* (motion detection).
when_no_motion	Function that is executed when the sensor changes state from *active* to *inactive* (no motion detection).

Wiring Diagram

Before moving on to the python code, it is necessary to assemble the electrical circuit that connects the PIR sensor to the Raspberry Pi. The figure below displays the necessary connections.

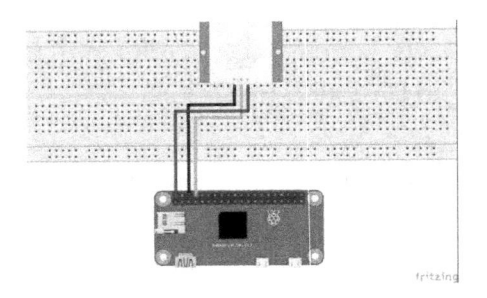

Connect the + and - pins of the sensor to *physical pin 1* and *3* of the *first GPIO line*, respectively. Connect the OUT pin of the sensor to BCM pin 14 (the fourth pin on the first GPIO line).

Detect Motion in Python

from gpiozero import MotionSensor
pir = MotionSensor(pin=14, pull_up=False)
print("Init PIR: waiting for motion...")
pir.wait_for_motion()
print("WARNING: motion detected!")

In this first usage example, we see how to use the constructor. We set the pin on which we connected the sensor output and the *pull-down* resistor (*pull_up=False*). After initializing the pir object we wait for motion to be detected by using the *wait_for_motion* method. Since we have not passed any *timeout* the PIR remains waiting for motion indefinitely.

```
from gpiozero import MotionSensor
pir = MotionSensor(pin=14, pull_up=False)
print("Init PIR: waiting for motion...")
while True:
pir.wait_for_motion(timeout=10)
if pir.is_active:
print("WARNING: motion detected!")
else:
print("no motion detected!")
```

In this second example, we implement a more sophisticated motion detection. This time we pass a timeout of 10 seconds to *the wait_for_motion* method as a parameter. This means that the method waits for at most 10 seconds for movement and then returns control to the main stream. In case of movement, or end of timeout, we check the value of the *is_active* property. If it is *True* "motion has been detected", if it is *False* it means "*no motion has been detected.*" In both cases, we convey an appropriate message. At this point, the loop starts again waiting for motion.

```
from gpiozero import MotionSensor
pir = MotionSensor(pin=14, pull_up=False)

from gpiozero import MotionSensor
def motion_alert():
print("Sending ALARM message: Motion detected!")
pir = MotionSensor(pin=14, pull_up=False)
pir.when_motion = motion_alert
print("Init PIR: waiting for motion...")
while True:
pir.wait_for_motion()
```

In this last example, we use the event mechanism by registering a handler called *motion_alert*. Then, after creating the pir object, we register the handler via *pir.when_motion*. Then we place ourselves on hold. In case the PIR sensor detects a movement the motion event is generated and its handler (*motion_alert*) is immediately invoked.

In this chapter, we have covered the use of a PIR sensor. GPIOZero provides a high-level abstraction making motion detection extremely intuitive. In addition to the wiring diagram some examples of usage in python and GPIOZero were analyzed. This sensor can be used in many projects (for example a home burglar alarm or a robot that reacts to the presence of external stresses).

Chapter 24

LDR Light sensor with GPIOZero

In this chapter, we will experiment with detecting ambient light change using a *Light Dependent Resistor* (*LDR*) or Photo Resistor and using the GPIOZero library.

Photoresistor

A photoresistor (LDR) is a device that varies its resistance as the incident light changes. In other words, it is a variable resistance according to the light intensity that is captured by it. The photoresistor is the basis of the operation of the classic twilight switch that is found on the market.

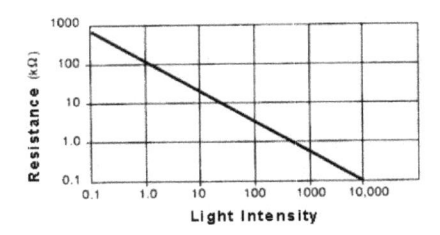

As you can see in the graph above an LDR has *low resistance* in the presence of light (order of K Ohm) while in dark conditions it assumes a very high value (order of Mega Ohm).

Circuit

Let's now see the connections to be made and introduce a little trick to use the photoresistor. Our Raspberry has only digital inputs and is not able to read analog values. To overcome this problem we will use a small capacitor together with the LDR.

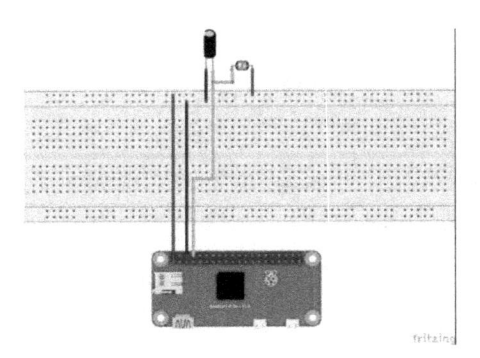

The capacitor in question is a *1 uF* electrolytic. We connect the negative *rheophore* of the capacitor to the GND pin of the board through the use of a breadboard. We connect one pin of the LDR to VCC (3.3V) provided by the Raspberry. Now we connect the positive pin of the capacitor with the free pin of the LDR. Both should be connected to GPIO 14 as in the figure. In this way, we have made a simple RC (*Resistor-Capacitor*) network.

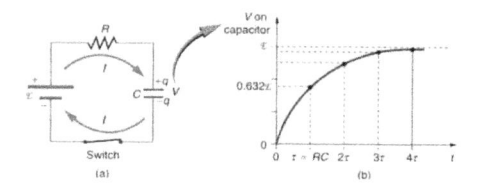

As can be seen in the figure, once a constant voltage is applied, the capacitor begins to charge exponentially. A fundamental parameter is the time constant which is equivalent to the product of resistance R by capacitance C. So this means that fixed capacitance C (1 uF capacitor) the charge time of the capacitor will be directly proportional to the value of resistance R. Now since the resistance value of the LDR, as seen before, is proportional to the light intensity, we can conclude that the charge time of the capacitor is directly proportional to the amount of light. Therefore, using this concept, the library GPIOZero measures the charge time of the capacitor being able to read an analog value in an indirect way. In other words, if the charge time is less than a certain threshold (10 mS) then the light condition is detected. On the contrary, if the charge takes more than 10mS then the absence of light or darkness will be detected. Let's see now the control code.

LightSensor Class

The LDR photoresistor, within the GPIOZero library, belongs to the input device class (*SmoothedInputDevice*). The constructor of the class has the following syntax:

LightSensor(pin=None,
queue_len=5,
charge_time_limit=0.01,
threshold=0.1,
partial=False,
pin_factory=None)

- *pin*: it's a mandatory integer parameter that represents the BCM pin number to which the LDR sensor is connected;
- *queue_len*: it's an optional integer parameter that represents the length of the queue used to store the past readings of the sensor. By default, it is set to 5. In case the LDR sensor is unstable it may be useful to increase the value of this parameter;
- *charge_time_limit*: it's an optional parameter of float type. It represents the charge time beyond which an absence of light is considered. The default value 10mS (0.01 S) is appropriate for a 1 uF capacitor. If a capacitor of different capacitance is used, it may be necessary to vary this parameter, but otherwise, it is generally not necessary to change it;
- *threshold*: it's an optional float parameter that represents the threshold, calculated on the average of the values present in the internal queue, beyond which the sensor is considered active. Normally it is not necessary to modify this parameter;
- *partial*: it's an optional boolean parameter set by default to *False*. When set to False the value of the *is_active* property is returned only if the internal queue is full (blocking reading). If it is set to True the reading is non-blocking, therefore the is_active value is calculated only

with the values present in the internal queue without waiting for it to fill up;
- *pin_factory*: it's an optional parameter for advanced use. It allows to specify a GPIO pin factory different from the one provided by GPIOZero.

The most important methods and properties of the class are as follows:

wait_for_dark(timeout=None)	It suspends the running instruction stream until the sensor detects no light (dark) or the timeout has expired.
wait_for_light(timeout=None)	It suspends the running instruction stream until the sensor detects light or the timeout has expired.
when_dark	Function that is executed when the sensor detects the absence of light.
when_light	Function that is executed when the sensor detects the presence of light.

Light Sensor

from gpiozero import LightSensor
ldr = LightSensor(14)
print("Active LDR sensor...")
ldr.wait_for_light()
print("Light Detected!")

In this first example, we implement a simple light detector. We instantiate the *LightSensor* class specifying the GPIO 14 on which the LDR is connected. Once the sensor is activated we wait for light by invoking the *wait_for_light* method which pauses the script until "light" is detected by the sensor.

Darkness Sensor

```
from gpiozero import LightSensor
ldr = LightSensor(14)
print("Active LDR sensor...")
ldr.wait_for_dark()
print("Darkness Detected!")
```

This example is similar to the previous one but we invoke the *wait_for_dark* method.

Twilight switch

```
from gpiozero import LightSensor
from signal import pause
def darkness_action():
print("Darkness detected!")
def light_action():
print("Light detected!")
ldr = LightSensor(14)
ldr.when_dark = darkness_action
ldr.when_light = light_action
print("Active LDR sensor...")
pause()
```

In this example, we realize the classic operation scheme of a twilight switch. Through the methods *when_dark* and *when_light* we specify two functions to be executed respectively when darkness is detected and when light is detected. In this way, for example, we can drive a relay that when it gets dark activates a lamp to illuminate a room. On the contrary, when the light rises, it closes the relay turning off the lamp.

In this chapter, we saw how, with a little trickery, we can exploit an analog LDR sensor to detect the presence or absence of light. Through the use of the GPIOZero library we realized a simple light and dark sensor. Finally, we have implemented the scheme of the classic twilight sensor.

Chapter 25

Raspberry PI 4: More Power, More RAM and 4K Video

Starting at just $35 for the 2GB model, the Raspberry Pi 4 is the world's best single-board computer, a must-have for tech enthusiasts of all ages.

Raspberry Pi has long been the gold standard for low-cost single-board computing, powering everything from robots to smart home devices to digital kiosks. When it launched in 2019, the Raspberry Pi 4 demonstrated good enough performance to be used as a desktop PC, as well as the ability to output 4K video at 60 Hz or power dual monitors. More recently, the Raspberry Pi 4 (8GB) model came out, offering enough RAM for serious desktop computing, productivity, and database hosting.

The Raspberry Pi 4 can serve as an educational PC for kids, a media center, a web server, a game emulation machine, or as the brain of a robot or IoT device. It opens up a whole world of possibilities for improving your life and having fun.

In this Raspberry Pi 4 chapter, we will help you answer the key questions you need to choose the right Raspberry Pi 4 model and make the most of one if you already own one.

Should I Buy a Raspberry Pi 4?

We've extensively outlined why every tech geek should own a Raspberry Pi. But choosing which Raspberry Pi to buy is an open question because there are reasons why you might want a different model, such as the $5 Raspberry Pi Zero. However, if you're looking for an all-around, generic Raspberry Pi, there's no question that it's the Raspberry Pi 4, which goes by the official model name Raspberry Pi 4 B (there's no 4 A, so the B is redundant).

How Does Raspberry Pi 4 Improve Over Other Models?

The biggest changes are the faster processor, a 1.5 GHz Broadcom CPU and GPU, increasingly faster RAM, the addition of USB 3 ports, dual micro HDMI ports (instead of a single HDMI connection), and support for 4K output. The increased bus speed that enables USB 3 support also allows the on-board Ethernet port to support true Gigabit connections (125 MBps) where last generation models had a theoretical maximum of only 41 MBps. The microSD card slot is also twice as fast, offering a theoretical maximum of 50 MBps versus the 3B+'s 25 MBps.

Because the new SoC needs more power, the Raspberry Pi 4 B charges via USB Type-C instead of micro USB. It also requires a power adapter capable of delivering at least 3 amps of power and 5 volts, though you can get away with 2.5 amps if you don't connect many peripherals to the USB ports. Putting the power requirements aside, the USB Type-C connectors are reversible, making them much easier for kids (and adults) to plug in.

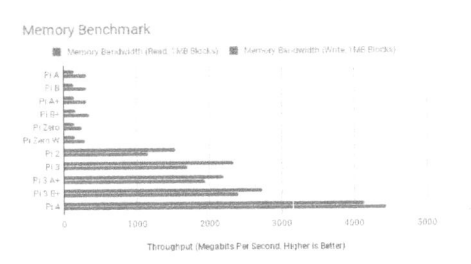

Memory Benchmark

Throughput (Megabits Per Second, Higher is Better)

The Raspberry Pi 4 has a similar design and size to its predecessors, but it's a completely new platform, powered by a new processor, the Broadcom BCM2711B0. Since the first Pi in 2012, all Pis have used 40nm SoCs, but this new chip is based on a 28nm process and, instead of the old Cortex-A53 microarchitecture, uses Cortex-A72. The BCM2711B0 in the Raspberry Pi 4 has four CPU cores and has a clock speed of 1.5 GHz, which at first glance doesn't seem much faster than the quad-core, 1.4-GHz BCM2837B0 in the Raspberry Pi 3B+.

However, the Cortex A72 has a pipeline depth of 15 instructions, compared to only 8 on its predecessor, and it also provides non-standard execution, so it doesn't wait for the output of one process to start another. So even at the same clock speed (and the BCM2711B0 is based on a smaller process node), Cortex-A72 processors will be significantly faster and use more power than their A53-powered ancestors.

For example, on the Linpack benchmark, which measures overall computing power, the Pi 4 absolutely beat the Pi 3 B+ in all three tests. In the single-precision (SP) test, the Pi 4 scored 925 points compared to the Pi 3 B+'s 224, a 413% increase.

RAM is also a bit faster, going from 1GB of DDR2 RAM operating on the Pi 3B+ up to 8GB of DDR4 RAM on the Pi 4. In addition to the increase in bandwidth, having more memory is huge, especially for web browsing.

The GPU also got a nice boost, going from a Broadcom VideoCore IV running at a core clock speed of 400 MHz on the Pi 3 B to a VideoCore VI set at 500 MHz. The new architecture allows it to output to a display with up to 4K resolution at a rate of 60 frames per second or support dual monitors at up to 4K 30 Hz.

Which Raspberry Pi 4 Should I Buy?

There are three current models of the Raspberry Pi 4 that are identical except for the amount of RAM. For $35, the entry-level model has 2GB of RAM, which is enough for most projects, from robots to retro arcade machines, but if you're using the Raspberry Pi 4 as a desktop PC, you should get the 4GB model, which is worth $55.

The official Raspberry Pi operating system (formerly known as Raspbian) is so memory efficient that we found it difficult to exceed 4GB, even with a ton of browser tabs open, videos playing, and several applications running. However, as more applications come out that take advantage of it, the $75 Raspberry Pi 4 (8GB) model will have more utility. If you can afford the extra $20 over the 4GB model, it's a good future-proofing idea.

Should I Get a Case for the Raspberry Pi 4?

At 3.5 x 2.3 x 0.76 inches (88 x 58 x 19.5 mm) and 0.1 pounds (46 g), the Pi 4 is thin enough to fit in your pocket and light enough to carry anywhere. The board is durable enough to probably survive rolling around in your bag, but if you're moving it around a lot, we recommend stuffing it into a case, especially to protect the pins. However, I often use the board naked on my desk and have even slipped it into a backpack pocket without incident.

If you want a case, be sure to choose one designed for the Raspberry Pi 4 (cases for earlier models will not fit). We recommend buying a case that leaves the GPIO pins accessible.

What Ports Does the Raspberry Pi 4 Have?

The Raspberry Pi 4 doesn't just cover the basics when it comes to ports. The right side has four USB Type-A connections, two of which are USB 3.0. There's also a full-sized Gigabit Ethernet port there for wired connections. The bottom edge has a 3.5mm audio jack, two micro HDMI ports, and the USB Type-C charging port. On the left side is the microSD card reader.

And on the top surface of the board, you'll see ribbon connectors for the camera serial interface (CSI) and display serial interface (DSI), which provide dedicated connections to the Raspberry Pi's camera and display (or compatible accessories).

There are many things you can do with the CSI port, including using a Raspberry Pi Camera as a PC webcam or turning it into a motion alert security camera. Of course, you can also connect a camera to a USB port, and there are a couple of more common ways, including micro HDMI ports, for output to a screen.

Perhaps the most important interface on any Raspberry Pi is its set of GPIO pins. With these, you can connect to lights, motors, sensors, and a huge ecosystem of HATs, which are expansion boards that attach to the top of the Pi. See the GPIO section below for more details.

What Kind of Power Adapter Do I Need for the Raspberry Pi 4?

You'll need a power source that can deliver at least 3 amps and 5 volts on a USB Type-C cable. The official Raspberry Pi 4 power supply, which costs about $10, does the trick, but so will any phone or laptop charger that meets these minimum standards and outputs on USB-C. You can also power the Pi 4 from a USB PD power bank that you would use to charge a phone.

Depending on how much power the PC can produce, you may be able to power a Raspberry Pi 4 from its USB-C port, though you may see a lightning bolt icon appear in the upper right corner of the screen, meaning the board is running on reduced power.

Like every Raspberry Pi model ever made, the Raspberry Pi 4 doesn't have a power switch. The default way to turn on a Raspberry Pi is to simply plug it in. When you're ready to turn it off, you turn off the operating system and then unplug the cable. You can also buy power switches that turn the power on and off, but don't forget to turn off the operating system before cutting the electricity.

Note that all previous versions of Raspberry Pi used micro USB connectors for power and could work with a 5 volt, 2.5 amp (or often lower) power supply. So, if you have a power adapter from a Raspberry Pi 3, it will not work with your Raspberry Pi 4.

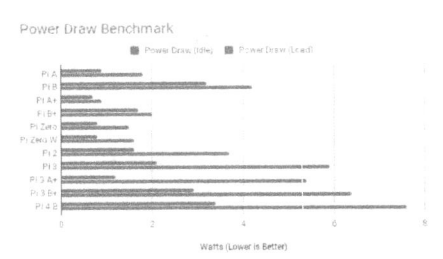

Which USB Type-C Cables Work With Raspberry Pi 4?

In theory, any Type C to Type C USB cable should work, but models of the Raspberry Pi 4 produced before the early 2020s had a small bug that prevented them from charging to "e-marked" USB cables. E-marked USB cables are usually the ones that offer high-speed data transfer over USB 3 at 10 Gbps. On the other hand, any cable that is USB 2.0 will definitely work with the Pi 4, as will many 5 Gbps USB 3.1 cables.

I tested a number of USB-C cables on a Raspberry Pi 4 and found that the vast majority worked, with the main exceptions an Apple MacBook charging cable and two 10 Gbps cables. Considering you're only using this wire for charging (the USB-C port only accepts power), there's no reason to get one that supports high-speed data transfer. Raspberry Pi 4s that were produced in the early 2020s and later have this issue fixed.

Chapter 26

20 Raspberry Pi Project Ideas

The fields of application for the Raspberry Pi span all areas. In addition to the many typical uses for which the miniature computer would seem predestined, there were also some amazing ideas presented that could be realized with the Raspberry Pi. Sometimes these projects require little knowledge, other times a lot of knowledge is needed, but with the right amount of interest, you will still be able to use it for your own projects. In fact, experimenting with the board and learning new knowledge in computing is the basic idea from which the computer was born.

The web is full of information for the creation of countless applications for the Raspberry Pi. The following examples are just to get a rough idea of the possibilities offered by the mini PC. I will provide a brief guide for each project inviting you to search for more detailed information on the project that interests you the most.

Web Server

Many users use the Raspberry Pi as a web server, including for example **Apache**, including lighttpd or NGINX. But to smoothly manage complex and dynamic websites, the Raspberry Pi's performance is not enough. The small computer is best suited as a local test environment for the site, but you can also host simple pages where you don't expect a large influx of visitors.

Smart Coffee Machine Pump Controlled by Raspberry Pi & HC-SR04 Ultrasonic Sensor

Coffee machines are almost indispensable at home and in the office. With modern devices, the beloved wake-up drink can be prepared at the touch of a button, provided the integrated water container is full. Alex Stakhanov and his colleagues added an automatic pumping system to their company machine, a SAECO Aulika Focus, which they built themselves, based primarily on a Raspberry Pi. An ultrasonic sensor HC-SR04, connected to the mini-computer, regularly measures the water level, while a software programmed with Python takes care of the necessary operation of the system. You can find a detailed explanation about this project made with the Raspberry Pi on the medium.com.

IOT

Increasingly popular is the connection of all home appliances. Smart Home, which allows centralized management and control of radiators, lamps, blinds, refrigerator, washing machine, and other appliances, not only increases the quality of life and comfort at home but also helps to achieve more efficient energy consumption. Due to its affordable price, ability to connect to the Internet, and its status as a standalone system with excellent hardware, more and more enthusiasts are using a Raspberry Pi to implement similar projects for their homes. The necessary software base is open source tools such as openHAB or Home Assistant.

Mail Server

If you use the Raspberry Pi as a mail server, your emails will be saved exclusively on this computer, as no other provider or server has access to your messages. With your own mail server you not only have complete control over your system but with your own domain, you can also create as many email addresses as you want. In this way, the mini-computer, as a central platform for the exchange of electronic messages, offers you the best security of privacy (since all data are under your control), but also a high flexibility.

Security System

With a Raspberry Pi you can get more comfort, but also more security within your four walls. Max Williams has in fact used the mini computer (Raspberry Pi 3 Model A+) as the basis for a small but sophisticated security system, which, once activated, scans the surrounding environment in real-time and in case of motion detection immediately sends a Telegram message including photos. The message is sent even if the device is activated or deactivated.

LED Interactive Interface

Do-it-yourself enthusiast Vincent Deconinck has demonstrated that the Raspberry Pi is not only interesting for projects with practical uses, in fact for about 130 dollars he has equipped an ordinary IKEA table with an interactive display, which reacts to objects placed on it with colorful animations and even allows you to play Tetris. The heart of the project is a Raspberry Pi, which processes all interactions recorded via an Arduino microcontroller and transforms them into the relevant animations thanks to the Glediator software.

Smart LED Window

The combination of LED units and the Raspberry Pi is not only limited to recreational projects: user *dannyk6* has published on *instructables.com* a guide to building a practical LED window that simulates sunlight. Windowless rooms and basements are thus bathed in a new light and thus convey a new, complete basic atmosphere. The mock window can be controlled via a web interface, where the brightness is set manually or adjusted automatically based on time and weather (via a Yahoo! API).

VPN Server

With a VPN (Virtual Private Network) you can encrypt data traffic in a network, which is especially recommended if you use public Wi-Fi; without any encryption, it is theoretically possible to intercept sensitive data at any time. This is why a VPN server is a great help, as it works quite easily on a Raspberry Pi. As a central authentication and broadcast instance for individual VPN clients requesting access, the mini-computer makes a good impression in both a private and corporate setting.

Binary Clock

If you've always dreamed of having a binary clock, Simon Monk's Raspberry Pi project is just the thing for you. The developer and writer has equipped the mini-computer with a Unicorn Hat, an add-on board with 64 RGB LEDs, which presents the current time in binary code thanks to a special software. Starting from top to bottom, this special clock provides the year (the last two digits), month, day, hour (in 24-hour format), minutes, seconds, and even hundredths of a second. A detailed guide was published in the forty-second issue of The MapPi magazine, which can also be found in abridged form on the official Raspberry Pi website.

Ted - the Talking Toaster

Voice control is one of the most important topics in the recent history of technology. What is therefore in favor of the creation of a talking toaster that reacts to voice commands, has been well exposed by the duo of developers "*8 Bits and a Byte*" that with Ted have launched this invention on the market. Although in the case of this project made with the Raspberry Pi the playful element is in the foreground, the fun device is a clear example of the possibilities and flexibility offered by the mini-computer. The toaster's voice functions are based on the Google AIY Voice Kit, while a Raspberry Pi 3 Model B (with a camera module included) takes care of the necessary computing power. In the post "*Ted the talking Toaster*" on *instructables.com* you can find detailed information about the project.

DNS Server

Thanks to a DNS server the resolution of the domain name into an IP address takes place. You can speed up this process in a local network by setting up your own DNS server on the Raspberry Pi. An own DNS server brings with it many other advantages: speed, security, ad blocking, child protection, etc.

AirPi Mobile Weather and Air Pressure Measuring Station

"The AirPi is a Raspberry Pi shield kit, capable of recording and uploading information about temperature, humidity, air pressure, light levels, UV levels, carbon monoxide, nitrogen dioxide and smoke level to the internet."

Air pollution can quickly become a problem that puts your health at risk. Obtaining reliable values of the air quality in your environment is often very difficult. A solution is offered by "**AirPi**": the package, consisting of Raspberry Pi and several sensors, allows the measurement of values, such as temperature, humidity, pressure, UV index, levels of monoxide, and nitrogen dioxide. In addition to air quality information, the Raspberry Pi application also provides weather information. A web interface allows you to view the measured values through the mini computer's Internet connection.

ownCloud

With the small computer, you also have a private cloud service, thanks to the free ownCloud software. Again, the Raspberry Pi acts as a server, where you can upload your data and use it when required. As opposed to commercial storage services, such as Dropbox or iCloud, by running your own cloud server you will have the great advantage of benefiting from complete control over your server and stored data, so you can safely save even sensitive data. In this article, understand how exactly it works and what other benefits ownCloud management brings (such as how to log in with the app).

Advice Machine

Receiving good advice doesn't always have to be expensive, as Nick Johnson's Advice Machine demonstrates. The Advice Machine, operated by a Raspberry Pi, has stored famous sayings that it provides for a fee. The quality of the advice, issued in the form of receipts thanks to a thermographic printer, varies depending on the number of coins entered. The input for these tips, popular sayings, and jokes, not to be taken too seriously, comes from the Fortune database, which traditionally entertains based on Unix and Linux systems.

Home Server e Media center

Those who want to make their home data available on all devices can rely on a home server. A home server is a file server on which you can store all kinds of data (documents, images, videos, music, etc.) and have access to the devices connected to the server (PCs, notebooks, smartphones, tablets, etc.). The connection is made via cable or Wi-Fi.

But that's not all: you can also go further and use the Raspberry Pi as a media center. So with the mini PC you can play music, movies, and images stored on the hard drive, but also use streaming services, such as the various online media libraries, YouTube or Spotify. A very popular software to configure a Raspberry Pi as a media center is **Kodi**, where all media files are sorted by type and described through illustrations.

Multiple Raspberry PI 3D Scanner

A very expensive but impressive project, costing around 11,000 euros, is the Pi 3D scanner, developed and perfected last year by Dutchman *Richard Garsthagen*. The basic structure of this two-meter-high body scanner consists of 100 Raspberry Pi's, each including its own SD card and camera modules. Using self-programmed 3D scanner management software, the recorded values can be optimized and used to print a 3D model.

Video Game Consoles

The performance of the Raspberry Pi is more than enough to reproduce old arcade video games or old consoles (via emulators). Enthusiasts have rebuilt arcade video games both in a miniature version and in the original size (or nearly so), even giving the possibility to insert tokens, thus recreating the authentic experience of these stations. As a software base for similar projects made with the Raspberry Pi, the combination created with the Raspbian and the **RetroPie** emulator is particularly popular.

MagicMirror

MagicMirror is a project for Raspberry Pi invented by Dutchman Michael Teeuw. It is a one-way mirror behind which a monitor and the small computer are mounted, while on the glass of the mirror are shown the time, weather, scheduled appointments, and much more. Also because of the great success that the developers have achieved after the publication of the DIY guide, there is already a second optimized version of MagicMirror, which thanks to its modular structure can be expanded almost without limits. Since the code is entirely open source, the magic mirror has developed over the previous years and given rise to a huge community, where the magicmirror.builders page serves as an exchange platform and reference point for interested parties.

Zelda Home Automation

A special kind of project made with the Raspberry Pi came to the mind of YouTuber Allen Pan aka *Sufficiently Advanced*: he, too, created a central instance for the Smart Home for easy control of technical devices within the home, managed by a Raspberry Pi. The operation of this home automation system, however, is not via voice command, text commands, or a web interface, but with the playback of melodies from a Nintendo classic "*The Legend of Zelda: Ocarina of Time.*" Like Link, the pioneering video game hero of his time, the YouTube star uses an ocarina (a wind instrument) to best play "Zelda's Lullaby" to open the front door, for example.

Siri: Open my Garage Door!

Already at the end of 2012, the user *DarkTherapy* showed how it is possible to open a garage in a few steps using a practical voice command. In an article on the official Raspberry Pi forum he explained how he converted his iPhone into a perfect remote control, using a Raspberry Pi and SiriProxy software. As an instance of voice recording on the Apple device was used the voice recognition software Siri, installed by default.

Conclusion

This chapter concludes our journey to discover the programming of the Raspberry Pi and its connector GPIO. In the guide we have analyzed, both theoretically and with practical examples, the use of the main classes of GPIOZero library. We have also interfaced input sensors like a push button, PIR presence detector, output actuators like relay and led. Through the various lessons, we have also learned how the GPIOZero library provides a clear and easy to use programming interface. My suggestion to you is to keep experimenting and make projects with two, three, or even more sensors and actuators. Starting from the examples discussed in this guide, the reader can develop his skills by moving on to more complex projects.

Enjoy experimenting!